Priority Technologies

The MIT Press's publishing mission benefits from the generosity of our donors, including Hyun-A Park and Jacob Friis.

Priority Technologies

Ensuring US Security and Shared Prosperity

edited by Elisabeth B. Reynolds

foreword by Simon Johnson

The MIT Press

Cambridge, Massachusetts | London, England

The MIT Press
Massachusetts Institute of Technology
77 Massachusetts Avenue
Cambridge, MA 02139
mitpress.mit.edu

The MIT Press would like to thank the anonymous peer reviewers who provided comments on drafts of this book. The generous work of academic experts is essential for establishing the authority and quality of our publications. We acknowledge with gratitude the contributions of these otherwise uncredited readers.

This book was set in ITC Stone Serif Std and ITC Stone Sans Std by New Best-set Typesetters Ltd. Printed and bound in the United States of America.

Library of Congress Cataloging-in-Publication Data

Names: Reynolds, Elisabeth B. editor | Johnson, Simon, 1963– writer of foreword
Title: Priority technologies : ensuring US security and shared prosperity / edited by Elisabeth B. Reynolds ; foreword by Simon Johnson.
Description: Cambridge, Massachusetts : The MIT Press, [2026] | Includes bibliographical references and index.
Identifiers: LCCN 2025048304 (print) | LCCN 2025048305 (ebook) | ISBN 9780262054294 hardcover | ISBN 9780262054300 pdf | ISBN 9780262054317 epub
Subjects: LCSH: Technology and state—United States | National security—United States | Economic security—United States
Classification: LCC T21 .P75 2026 (print) | LCC T21 (ebook)
LC record available at https://lccn.loc.gov/2025048304
LC ebook record available at https://lccn.loc.gov/2025048305

10 9 8 7 6 5 4 3 2 1

EU Authorised Representative: Easy Access System Europe, Mustamäe tee 50, 10621 Tallinn, Estonia | Email: gpsr.requests@easproject.com

Contents

Foreword

In 1932, British Prime Minister Stanley Baldwin delivered a memorable speech in the House of Commons, addressing head-on the question of how rapidly improving technology would change the nature of war. In Baldwin's view, the risk of mass death from enemy bombing was now very real, and for a simple reason:

> the appalling speed which the air has brought into modern warfare; the speed of the attack. The speed of the attack, compared with the attack of an army, is as the speed of a motorcar to that of a four-in-hand.[1]

In 1918, the fastest heavy bomber in the world was arguably a Handley Page twin-engine biplane with a top speed of 97 mph. By 1932, the latest bombers (some of which were still prototypes) could already reach close to 200 mph. Such fast-moving bombers meant that

> In the next war you will find that any town within reach of an aerodrome can be bombed within the first five minutes of war to an extent inconceivable in the last War, and the question is, Whose morale will be shattered quickest by that preliminary bombing?

This thought had profound implications—for Baldwin, for the British government, and for militaries everywhere—in terms of thinking about what mattered. Baldwin put it this way:

> I think it is well also for the man in the street to realize that there is no power on earth that can protect him from being bombed, whatever people may tell him. The bomber will always get through.

The implication was simple: build bigger, faster bombers. And one prominent strategic thought—which has been with us ever since—was: Bomb the enemy early in the next war. Perhaps even bomb the enemy's airfields and destroy as many opposing aircraft as possible on the ground before formally declaring war. Exactly the same thought was conveyed brilliantly and reinforced in dramatic fashion by H. G. Wells in *The Shape of Things to Come*, published in 1933. (The book became a compelling 1936 movie about the imminent future of war.)

In retrospect, Prime Minister Baldwin was spot on in terms of the continued technological improvement in bombers—in 1940, the long-range German bomber, the Heinkel He 111, had a top speed of around 270 mph. And Baldwin was also in the right ballpark regarding the ultimate long-run implications of increased bombing capacity—by 1943, the Allies had launched an air campaign that devastated German industry.

Baldwin was channeling the thinking of leading air force officers and theorists (and H. G. Wells himself, who had been writing about air power for decades). But Baldwin and almost all of these military experts and visionaries got one big thing wrong: "The bomber will always get through" was not correct. By the end of the 1930s, effective air defense against bombers was not only a real possibility—for Great Britain, at least, it had become the decisive weapon. In the Battle of Britain, between July and October 1940, RAF Spitfires and Hurricanes turned back German daylight bombing raids and effectively prevented an invasion. German bombing did plenty of damage, including to London and other cities during the Blitz. But the original goal of the German bombing campaign was to seize control of the air, making it possible for ships and ground forces to move with impunity. On this key dimension, the German bombing effort failed entirely.

The reason for this shift in the fate of bombing runs was simple. At the same time as bombers became faster, the speed and operational ceiling of fighter planes, a.k.a. interceptors (as they were designed to intercept bombers), also improved. Fighter planes at the end of World War I struggled to stop determined bombing attacks, and military experts soon dreamed that bombers would become flying fortresses that could defend themselves effectively. But by 1940, the Spitfire could fly at over 350 mph—and dive at perhaps 500 mph. Not only did the British have greatly improved fighter planes, including the Hurricane, but they also built a radar and human observer system that could pick up early warnings of incoming bombers, determine their likely vector, process this information in real time, and send guidance to fighter command centers—actionable intelligence indeed.[2]

Essentially, the British won the Battle of Britain by doing what Baldwin—and most other experts in 1932—thought was impossible. In effect, the RAF built a massive, human-powered analog computer that could determine where the daylight bombing threats were and dispatch fighter aircraft that could break up those attacks. Despite German countermeasures and the strong performance of their own fighter aircraft (particularly the Messerschmitt Bf 109), British air defense prevailed.

What are the lessons from that experience for the United States today?

Everyone surely agrees that, as technology changes, so does war. This has been evident at least since early in World War I when the machine gun, the tank, and most of all aircraft completely altered how nations fight.

But while many technologies change all the time—and perhaps at faster rate today than ever before—it is impossible to say what will be decisive at any moment in time.

We do know that building big stockpiles of weapons is unlikely to be effective, at least when the pace of technological change is rapid.

In the early 1920s, France had the largest air force in the world. But all of those planes were obsolete by the 1930s (if not before), and France's lack of good fighter planes proved a disastrous weakness when Germany invaded in May 1940.

Drones are surely an important part of future conflict. But what size of drones, controlled how, and used in what way? And couldn't anti-drone defenses also develop at a rapid pace? I have no idea what the answer will be. I have listened carefully to the world's experts on these questions, as reflected in this volume, and I have learned a great deal. But mostly what I know is simply that I want these experts to keep working on these important questions, I want this work to be backed with significant resources, and I want people with all kinds of ideas from the full range of relevant experiences—including from Ukraine since 2022—to contribute to this agenda in the United States and in Western Europe.

If we don't know exactly what the future global economy will look like, then we don't know which technologies will be decisive—but we do know that today's stock of knowledge and capabilities is not likely to be enough. We have to keep innovating.

The most important innovations for national security will surely be those that move other technologies forward—for example, semiconductors or advanced manufacturing. It's also appealing to push hard on commercial applications when we can imagine that related or subsequent developments could have defense implications. The Germans were forbidden from having military aircraft after World War I, but they made good progress with civilian aircraft—and in the 1930s, brought those ideas and the associated expertise into their military buildup.

And that's another key lesson from history. Invention requires inventors—people who devote years to developing technologies, even when the particular line they are pursuing has not yet taken hold. A country like the United States should be investing in breakthroughs wherever these are possible—for example, in biomanufacturing today. And, as David Baker (Chemistry Nobel Prize laureate in 2024) and I have argued, the best way to promote any scientific enterprise is to strengthen the pipeline of young talent entering labs and thinking hard about innovation.[3] In many fields, key inventions are produced and powered by graduate students and postdocs. Increasing the number of talented people entering all scientific fields needs to again become a pressing national priority.

If you invent the industries of the future, you will get more of the good jobs that are created—a point that Jonathan Gruber and I emphasized in *Jump-Starting America: How Breakthrough Science Can Revive Economic Growth and the American Dream* (Public Affairs, 1999), and one that was taken up by the bipartisan Chips and Science Act of 2022. But you also strengthen your scientific cadres—your brilliant young people (and the great older experts) who can both build what you want today and also pivot, if needed, to build something different but likely related when needed.

Before World War II, the US government spent almost nothing on research and development. The lessons of that war—including those from programs to develop radar, synthetic rubber, and penicillin, and of course, the Manhattan Project—convinced Congress and the broader political leadership that supporting basic science was essential to accelerate technological progress. This progress proved not just something that was nice to have but rather an essential part of postwar American prosperity—and national security.

But what should we develop? Where are the frontiers? What is the set of potential breakthroughs that are available today, even if we cannot yet see the full contours? What are the essential inputs, such as critical minerals, that we will need? Where does "advanced manufacturing" fit into the defense picture?

As an economist, I also have to ask the question: How much will any of these initiatives cost us? How large are the potential spillover benefits? And why isn't the private sector going to do this of its own accord?

And please remember also this essential lesson from what happened at Pearl Harbor on December 7, 1941. The US Navy already used radar for fire control on ships, the Battle of Britain had demonstrated the value of radar for air defense almost eighteen months previously, and the British gave the cavity magnetron (a key element for the future of radar) to the United States in September 1940. But on the morning of December 7, when two radar operators reported an incoming flight of likely enemy planes, the officer in charge of the command center told them they were hallucinating, and they should go have breakfast.

Ultimately, effective defense is not about the hardware or the software. It is about how you incorporate technology into your decision-making apparatus, so that accurate-enough signals trigger effective responses. Or, to put the same point more forcefully, it's never about the machines—it's always about how you use them. Nothing is more important than the vision we have for the future of technology.

At MIT, we do not have all the answers for the deep puzzle of how to boost shared prosperity while mounting an effective defense against all possible threats. But we have some pieces of the answer, and publishing this volume is intended to encourage others to put forward their ideas. It is also intended to help us, across all parts of MIT, think harder about what is needed and why the federal government should put taxpayer resources into particular efforts.

Technology has been central to society at least since the first human figured out how to sharpen a stone. How militaries are organized and equipped has always been a central feature of any civilization. But how we organize production more broadly is just as important for the nature of our society.

Today's world is highly uncertain. But other countries are investing heavily in upgrading their scientific capabilities, taking pages from the post-1945 American playbook. Some of these countries see themselves as racing against the United States to invent the future, partly for commercial purposes (and to grab more of those good jobs), but also likely with at least some potential military-related rationale.

The United States is an incredibly inventive place—surely the most creative place, over the past 150 years and still today, in terms of imagining and implementing new technology. We have a population of close to 340 million. But the world has more than 8 billion people, most of whom would welcome better technologically powered solutions for a myriad set of problems. As Vannevar Bush wrote in his famous 1945 report, science really does offer us an endless frontier.

We can invent better futures for ourselves and for the world. We can do this in ways that are responsible, respectful to the environment, and careful with regard to all forms of unintended consequences (including what happens to jobs)—all points that Daron Acemoglu and I emphasized in our book, *Power and Progress: Our Thousand-Year Struggle Over Technology and Prosperity* (Public Affairs, 2023). And while we focus on this, we can also make ourselves, our allies, and the entire world a safer place.

To do this, we should invent, develop, and implement everything discussed and proposed in this volume. And a lot more.

Simon Johnson
December 2025

Notes

1. A "four-in-hand" was a horse-drawn carriage that could travel fast, for its day. For the full text, see Stanley Baldwin, "A Fear for the Future" (speech, House of Commons of the United Kingdom, November 10, 1932), https://missilethreat.csis.org/wp-content/uploads/2020/09/A-Fear-for-the-Future.pdf.

2. The deep bunker headquarters of 11 Group, with primary responsibility for South East England, is in Uxbridge and well worth a visit; anyone who cares about

the future of technology should visit this compelling piece of strategic history. RAF Fighter Headquarters at Bentley Priory—the brains of the information processing operation—is also worth a visit, although sadly the underground bunker there was not preserved.

3. David Baker and Simon Johnson, "Will the US Continue to Dominate Science?" in *The Economic Consequences of the Second Trump Administration: A Preliminary Assessment*, ed. Gary Gensler, Simon Johnson, Ugo Panizza, and Beatrice Weder di Mauro (CEPR Press, 2025).

Introduction

In the twentieth century, the United States led the world technologically and benefited economically, militarily, and socially from the positive spillovers that came with that leadership. From airplanes to microchips, nuclear energy to biotechnology, and the internet to GPS, a wide range of inventions and innovation have driven the US economy and positioned the country as the global leader in industries important to the country's overall security. Today, however, the country's global technological leadership is no longer assured and is in fact slipping, challenged primarily by China. At the same time, US national and economic security interests are being threatened by geopolitical rivalries, supply chain vulnerabilities, and energy security concerns, among other challenges. A changing economic world order is emerging—more multipolar, fragmented, and protected—that is prompting new strategies in the United States to secure technological leadership and advantage in key industries. The essays in this volume outline strategies and pathways forward to help the country regain its edge and achieve a more secure and prosperous future.

"Priority technologies" refers to those technologies where the development of scientific and technological capabilities is considered strategically important for the country's national security. The definition of the latter has expanded over time beyond military threats from other states and national defense to include economic security to advance economic growth and prosperity and protect the economy from threats such as the disruption of key supply chains or the loss of critical advanced technological capabilities. Increasingly, national and economic security have merged such that, as many suggest, it is hard to know where an economic interest ends and a national security interest begins. For example, recent efforts to build up US domestic production of semiconductors in the face of supply chain vulnerabilities were driven by both their importance to the national defense and their role in key industries like automotive, where a shortage of semiconductors in 2021 due to the global pandemic contributed directly to a rise in auto prices and increased inflation.

These technologies are typically characterized by being at the frontier of scientific and technological innovation and thus not easily substituted by other technologies. They also tend to be dual use, with both defense and civilian applications. Finally, the technologies have a global reach, in which both international competition and cooperation elevate interest in their development. All six of the technology areas discussed in this volume align with these characteristics; few would question their importance to national and economic security. But they are by no means exhaustive. For example, nuclear energy or AI, both of which are addressed by MIT researchers elsewhere, could easily have been included. The technologies we focus on here have instead emerged through discussions at MIT that aligned perspectives on the urgent technological opportunities and challenges facing the United States with work being conducted by MIT researchers.

While the title of the book is *Priority Technologies*, the technologies are presented within the broader context of priority industries. In some cases, there is significant overlap between technology and industry, as with semiconductors, biomanufacturing, or quantum computing. In other cases, like critical minerals or advanced manufacturing, a suite of technologies are important to advance US capabilities and capacity in related industries. The chapters in this volume speak to strategies that both advance technology and grow globally competitive industries. The former might require a focus, for example, on scientific research or technology transfer, while the latter might involve accelerating technology adoption, strengthening supply chains, and expanding the talent pipeline. Regardless, these strategies go hand in hand.

Cross-Cutting Themes

Read as a whole, the chapters capture the unique aspects and the role key stakeholders play in both US innovation and industrial ecosystems. As one looks across the common themes that emerge, it is worth noting how important the federal government is to many of the strategies outlined here. Because the government can take a long-term view and invest with an eye toward social as opposed to private returns, it plays a critical role in providing R&D funds to develop next-generation and "leapfrog" technologies before they are attractive to the private sector or before a commercial market even exists, as in the case of quantum computing. It can also invest in test beds and large-scale demonstrations that help advance the scale-up process, as it has done in biomanufacturing, as well as support the development of key inputs to supply chains, such as in critical minerals. Just as on the supply side, the federal government can also play an essential role on the demand side by being a first customer to help pull a technology toward commercialization through procurement processes or advanced market commitments. It can also reduce the cost of capital through low-cost loans, grants, and favorable tax incentives. Finally, the federal government can shape technological trajectories through regulation by federal agencies, as in the case of drones and the regulation of unmanned aircraft systems (UAS) by the Federal Aviation Administration.

Likewise, universities are another key stakeholder in the country's innovation system. As recipients of over 50 percent of federal R&D funding, they are an essential engine to driving invention and innovation at the early stages of R&D. They are also one of the primary pathways for promising innovations in the lab to transition into commercially viable products through startups that spin out of the university. For example, much of the frontier innovation in the semiconductor and quantum computing industries today comes from entrepreneurs and startups born from universities. Of course, at their core, universities educate the next generation of scientific and engineering talent. US leadership in the industries discussed in this volume will depend upon a significant increase in the number of undergraduate and graduate students in STEM (science, technology, engineering, and mathematics) fields, among others. Finally, universities are often anchors in regional innovation ecosystems or "technology hubs" where the US develops specialization and competitive advantages in particular technologies, working with government, industry consortia, labor, and other nonprofits within a region.

Turning to cross-cutting themes, one of the most important that emerges in the volume is education and workforce development. It is worth discussing this topic in a bit more depth because of its profound implications for the country's long-term growth and shared prosperity.

The United States faces a fundamental challenge: a shortage of trained workers at every education level across these technology sectors. Multiple converging trends are creating a long-term demographic challenge. The rising ratio of retirees to working-age adults, falling fertility and labor participation rates, and radically reduced immigration are constraining the US labor supply for years to come. Meanwhile, the historical focus on four-year college education while neglecting vocational training and community college systems has left the United States short of skilled technician-level workers.

Promising investments in education and training are underway. Regional workforce training programs are expanding—evidenced by new semiconductor training programs at community colleges and four-year institutions. Schools are experimenting with new K–12 education methods and career pipeline programs. However, employers must also partner with training providers and offer quality jobs with attractive career prospects to attract and retain workers. And as several chapters underscore, skilled immigrants have played a crucial role in the development of US innovation capacity to date and must be part of the country's workforce development strategy going forward.

AI's impact on the US labor force remains uncertain, but it will affect every industry discussed in this volume. One positive scenario, sketched by MIT's David Autor, is that industry, education, and the public sector could deploy AI to enhance worker productivity through collaboration rather than pure automation, benefiting many workers while mitigating demographic and cost pressures. Yet technology has, over many decades, tended to exacerbate inequalities and favor highly educated workers. We cannot assume that this time will be different. As MIT's Work of the Future task force emphasized, stronger US labor market institutions will be essential for supporting workers through this technological transition. Successfully navigating these workforce challenges will be central to America's future competitiveness.

Several other cross-cutting themes emerge as well:

- *Supply chain resilience*. Supply chains critical to the country's national and economic security are vulnerable to geopolitical and other threats. Both companies and countries are investing in greater supply chain resilience through increased transparency, connectivity, and sourcing diversification and redundancy. For the United States, improving supply chain resilience requires investing in domestic production capacity to address security issues but also building greater resilience in regional and global supply chains.

- *Manufacturing capabilities and capacity*. Rebuilding US manufacturing capabilities and capacity is a foundational issue either directly or indirectly across all of the chapters. In terms of capabilities, increased technology adoption and digitalization, particularly among small and medium-size manufacturers, is an imperative to increase productivity among other goals. With respect to capacity, the US needs to expand its ability to produce at scale, which will benefit from building regional manufacturing ecosystems and "learning by building." This will be important, for example, as the US looks to increase its drone production from tens of thousands a year to several million.
- *Scale-up finance and demand-side strategies*. Capital-intensive technologies often require larger amounts of capital investment over longer time horizons to move from the pilot phase to commercial production. US capital markets have historically been interested in "asset light" investments and are thus less interested in providing the growth capital needed to help hardware-based technologies scale. There are signs, however, of increasing investor interest in physical industries as well as experimentation with the use of public funds to catalyze private investment. Equally importantly, federal procurement and support for scale up through demand-side policies can help validate new technologies and accelerate their commercialization.
- *Research and development*. US global leadership in key technologies and industries requires significant investment in scientific and technological breakthroughs if the United States is to lead

globally. The opportunity to leapfrog the current state of play across these technologies occurs only with long-term sustained investment and bold goals that necessarily involve government funding and university research.

Structure of the Volume

The six chapters are ordered to provide an introduction across three dimensions of priority technologies. The first two chapters, on critical minerals and semiconductors, provide a view on what could be considered foundational technologies for the US economy. Critical minerals are the raw materials that power vast swaths of the modern economy, from defense to energy to transportation. They are upstream in the value chain as are semiconductors, another foundational technology that is an important input to key industries discussed in this volume. Both technology areas have been subject to geopolitical disruptions in recent years, highlighting the importance of ensuring resilient supply chains.

The next two chapters discuss biomanufacturing and quantum computing, which could be considered frontier industries (though there are frontiers to be crossed in all of the chapters), where there is significant potential for achieving US global leadership. However, reaching this goal will require continued investment in scientific and engineering breakthroughs. In these areas, public-sector investment is particularly important to support R&D and also to catalyze investment by the private sector, which often prefers to avoid the risks inherent in basic science, initial applications, and scale up.

Finally, the chapters on drones and advanced manufacturing underscore the importance of more downstream inputs to the value chain, including digital manufacturing, to enable the capture of the full value of production to the United States. They also highlight the current challenges that exist in expanding US manufacturing capabilities and capacity. In the case of drones, this must be done largely from scratch since virtually all drone parts are made in China, though the core technologies were invented in the United States and elsewhere—a familiar pattern. The drone chapter provides a critical case of the challenges for the US industrial base and how it should intersect with the US innovation ecosystem. The advanced manufacturing chapter is in some ways a "closer" to all of the chapters, expanding on many of the themes in the previous five chapters where advances in innovation require a strong US industrial base to effectively capitalize on the full potential of investments in innovation.

The following summarizes each of the chapters.

Chapter 1, "Critical Minerals: Diversifying Supply Chains to Drive Resilience and Innovation," underscores a fundamental challenge for US industrial strategy: the nation's overwhelming dependence on foreign—particularly Chinese—sources for materials essential to advanced manufacturing, clean energy, and national defense. Elsa A. Olivetti shows how this dependence has created strategic vulnerabilities that market forces alone cannot resolve. Through a coordinated strategy spanning mineral extraction, processing, and downstream manufacturing, China has achieved dominance across entire value chains. In response, the chapter outlines a comprehensive five-point plan including investment in recycling infrastructure, coordinated R&D, reform in permitting, demand-side policies, and supply chain transparency. The chapter emphasizes how an effective US critical minerals strategy requires simultaneous government action across multiple agencies, timeframes, and market segments to rebuild capabilities that private actors cannot sustain independently, given the dynamics of global commodity markets.

In chapter 2, "Semiconductors: Rebuilding the US Semiconductor Ecosystem for Sustained Global Leadership," Jesús A. del Alamo writes that semiconductors are "the oxygen of modern human society." In this chapter, he examines how a critical technology's migration of manufacturing capacity overseas can create profound national security vulnerabilities and outlines recent steps by the federal government to address them. In 2022, the United States manufactured 0 percent of the world's most advanced chips despite inventing the technology, with 92 percent produced in Taiwan, just 100 miles from mainland China. While public- and private-sector investments in semiconductor manufacturing capacity and R&D in the United States are showing early signs of success, formidable challenges remain to the country's investing in the long game. The United States will need a thoughtful and deliberate follow-up strategy that strengthens the entire US semiconductor ecosystem to ensure sustained global leadership in both the design and manufacturing of advanced semiconductor chips.

In chapter 3, "Biomanufacturing: Transforming US Capabilities and Capacity to Accelerate the Domestic Bioeconomy," J. Christopher Love reveals a striking paradox at the heart of America's ambitions for the bioeconomy: While the United States still leads globally in biotechnological innovations, it fundamentally lacks the manufacturing infrastructure to capitalize on a sector that is poised to generate over $4 trillion of value worldwide in the coming decade. Despite pioneering advances in synthetic biology and bioengineering, US companies routinely seek overseas manufacturing capabilities and facilities to introduce and scale their biological innovations, creating dangerous dependencies and ceding manufacturing leadership to competitors like China. This chapter explores how emerging leapfrog technologies—including continuous manufacturing systems and modular small-footprint production facilities that could be 10 to 100 times cheaper and faster to deploy—offer a pathway to rebuild domestic biomanufacturing capacity. Love proposes a comprehensive national strategy centered on accelerating the implementation of advanced manufacturing technologies, regional innovation hubs, and targeted financial incentives to transform how biological products move from laboratory breakthroughs to market-ready goods while establishing a strong industrial base for US biomanufacturing.

In chapter 4, "Quantum Computing: Enabling Long-Term US Leadership Through a Sustained and Bold Science and Engineering Agenda," William D. Oliver and Jonathan Ruane outline the current quantum computing landscape, which has profound implications for national security and economic competitiveness. Working quantum computers exist, but achieving "quantum advantage"—the point where quantum systems outperform classical computers on practical problems—remains the critical goal that could unlock a multibillion-dollar global market by 2040. The United States currently leads in private investment and commercial hardware availability, but China's aggressive public funding and surging patent activity pose serious competitive threats. The authors make the case for a national strategy that involves patient, consistent, and sustained research funding by government, coupled with specific initiatives to foster thriving ecosystems, a prepared workforce, and a resilient supply chain. A key goal is to ensure the prevention of a "quantum winter" and secure long-term US leadership in this strategically vital technology.

In chapter 5, "Drones: Advancing US Innovation in Autonomy and Opportunities for Reindustrialization," Fiona E. Murray examines a set of connected technologies and capabilities where American innovation is currently constrained by limited domestic manufacturing capacity and supply chains. Recent battlefield experience in Ukraine demonstrates both the military potential of mass-produced consumer drones and the risks of supply chain dependence on strategic competitors. While the United States remains competitive in autonomy research (and in the most sophisticated drones), its lack of manufacturing in components such as motors, batteries, and inexpensive electronics limits its ability to meet domestic demand in both civilian and defense markets and, in turn, potentially stifles long-term innovation. In contrast, Chinese firms dominate production, global markets, and, increasingly, innovation in autonomy products and services. Murray recommends approaches that can harness the United States' strengths in software and autonomy while rebuilding the industrial base needed for scaled drone production.

Finally, chapter 6, "Advanced Manufacturing: Driving the Next Industrial Revolution," discusses how advanced manufacturing undergirds the country's national security, economic competitiveness, and technological leadership. In this chapter, I highlight how digital technologies—from AI and robotics to additive manufacturing—are reshaping production systems and helping to address critical vulnerabilities exposed by recent supply chain disruptions, geopolitical tensions, and energy demands. The chapter outlines the core challenges facing US manufacturers, particularly small and medium enterprises, while providing recommendations to accelerate digital transformation, improve productivity, and rebuild the industrial ecosystems essential for competing in critical technologies. The analysis reveals both the urgent need to revitalize the US industrial base and the pathways available to build a more productive, resilient, and sustainable manufacturing sector with quality jobs for the twenty-first century.

This volume was designed from the outset to be accessible as well as practical, with a high-level overview that focuses on near-term strategies that can position the United States for success. The book speaks to a wide audience—industry leaders, government officials, policymakers, researchers, and readers interested in US competitiveness in the context of a shifting global political and economic landscape. Each chapter provides an introduction to a technology with a clear-eyed assessment as to where the US stands in terms of its development and a pathway forward for achieving global leadership and security.

Across all technologies, the pathway forward requires long-term, sustained, and coordinated efforts to meet current challenges. With the merging of national and economic security concerns, innovation and industrial capacity become central to both geopolitical priorities and economic strength. In decades past, US technological strategy might have focused primarily on US innovation capacity through investment in R&D and the defense industrial base. While these are still crucial, the range of policies and investments required today span a much broader agenda that includes resilience across strategic supply chains at home and abroad, manufacturing capabilities and capacity to produce at scale, and workforce development to build a robust pipeline of skilled workers.

The chapters in this book provide examples of how the US has invented new technologies or industries and moved downstream production capabilities offshore, only to determine that the country is worse off for the loss of manufacturing and subsequent innovation—as in semiconductors and drones. These cases are cautionary tales for other frontier industries such as biomanufacturing, where US leadership in innovation needs to be matched by innovation and expansion of production capacity to lead in the myriad opportunities in the bioeconomy or else lose it to other regions in the world. Likewise in quantum computing, long-term, consistent investment in frontier science and engineering also requires attention and investment in a secure supply chain and manufacturing capabilities. The complexity of the critical minerals industry underscores the numerous ways domestic mining requires active investment and engagement across the value chain from R&D to demand-side strategies, all within the context of respecting environmental and community concerns. Finally, advanced manufacturing technologies, enabled by digital technologies including AI, are foundational to all of the above and need to be deployed at scale to improve production processes, build resilient supply chains, reduce emissions, and support the scale-up process.

Achieving the ambitious goals outlined in this volume will require collective agreement not only on which priority technologies to pursue but also on the need to broadly share the benefits from these investments. National and economic security goals can be achieved while also providing opportunity for the country as a whole. Many of the strategies touched upon in the chapters—expanding workforce development, increasing manufacturing, diffusing technology, scaling more companies, investing in regional clusters—support economic growth while creating direct and indirect benefits for workers, companies, and communities. This volume offers a way forward to increase US capabilities and capacity in strategic technologies at home, and in the process, strengthen US competitiveness and leadership abroad.

Elisabeth B. Reynolds

1 Critical Minerals: Diversifying Supply Chains to Drive Resilience and Innovation

Elsa A. Olivetti

Strategic Importance

US demand for critical minerals continues to expand. Lithium demand alone rose by nearly 30 percent between 2022 and 2025, yet the United States is far from able to meet such demand with domestic supplies.[1] This makes the United States reliant on supply chains for critical minerals that are concentrated in other countries—sometimes hostile or not easily accessible—thus creating a national security and economic vulnerability.

Critical minerals play an important role in nationally important industries such as automotive, aerospace, semiconductor, and defense applications. Energy technologies are also minerals-intensive; for example, solar panels, windmills, and batteries all require minerals such as lithium, cobalt, nickel, and graphite.

The House Select Committee on the Strategic Competition Between the United States and the Chinese Communist Party emphasized in a December 2023 report that "the U.S. is dangerously dependent on the PRC for critical mineral imports."[2] China maintains a singular global position in the supply chains for numerous critical minerals because of its dominance in mineral extraction, metallurgical processing, and recovery processes as well as intermediate steps along the value chain. China also maintains a dominant position in the data and intelligence required to locate minerals and to understand their supply chains (figure 1.1).

The lack of resilient supply chains for critical minerals can hinder the ability of the United States to remain competitive in both technology deployment and relevant innovations in the materials extraction, processing, and manufacturing sector. The federal government and industry have begun to develop and implement a comprehensive strategy for critical minerals, but gaps remain that require collaborative models to define problems and develop solutions across government, industry, and academia.

Current Landscape

The US Geological Survey (USGS) identified fifty critical minerals in a 2022 report.[3] The USGS found that the United States is entirely dependent on imports for 12 of the 50 critical minerals and is more than 50 percent import-reliant on an additional 29.[4] China and Canada supply the largest number of mineral commodities for which the United States is net import-reliant, followed by Germany and Brazil. China's role is particularly critical, as it is the primary US supplier of graphite, tantalum, and yttrium—minerals for which the United States is 100 percent import-reliant—as well as rare earth elements (with over 95 percent import reliance).[5]

China's dominant position, especially in the processing segment of the supply chain, enables it to exert both geopolitical and market power to influence global value chains. This dominance can take the form of export restrictions, such as the April 2025 limits on seven rare earth elements,[6] or even greater threats made later in the year by China, which can create bottlenecks and threaten downstream industries with severe consequences for national and

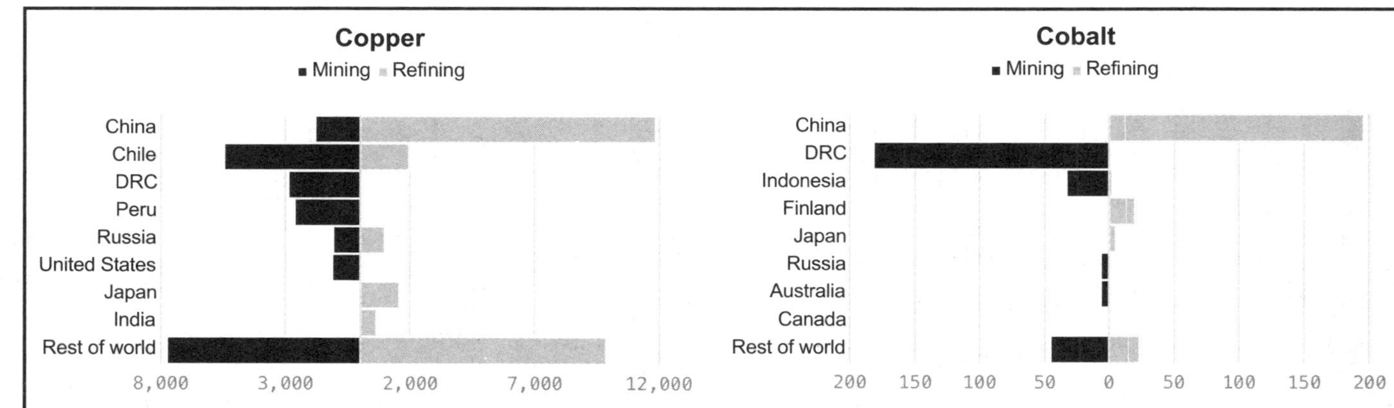

Figure 1.1

Global supply of copper and cobalt by country in 2024, showing both mining and refining capacity in kilotonnes. (DRC = Democratic Republic of Congo.) Complete data for all minerals is available in the appendix. Data from International Energy Agency, Global Critical Minerals Outlook 2025 (IEA, 2025), https://www.iea.org/reports/global-critical-minerals-outlook-2025.

economic security. Alternatively, China may oversupply the market to drive prices down, rendering operations in higher-cost countries (such as the United States or Australia) unprofitable. An example is the significant drop in lithium and nickel prices in 2022.

However, the USGS also points out that the United States has high potential critical mineral deposits,[7] especially in the western United States, and that many critical mineral deposits exist in allied counties. Moving from a mineral resource, defined as the estimated amount of minerals in a deposit, to a reserve, the portion deemed economically viable for extraction, is not primarily a matter of material availability, but of economic feasibility. The United States is not lacking in mineral occurrences, but without progress in market prices, extraction costs, technological innovation, and investment appetite, developing a domestic industry will remain a challenge.

Gaps and Opportunities

The role of the United States as a major consumer of critical minerals (both as a raw material and as part of finished products) is in some ways a potential competitive advantage. To the extent that critical materials can be obtained from recycling, the United States is well positioned, as it has plenty of products from which critical materials could be obtained. There is an opportunity for the United States to set preferences at a federal level for domestic sourcing and sustainability and to influence global supply chains through resulting customer activity. The federal government has already taken steps in 2025 to support more domestic sourcing through targeted investments and price floors for mining companies.

In addition, opportunities for mineral extraction in the United States may increase because ore quality is declining worldwide. This can make US reserves more economically attractive if US extraction and processing become state of the art and achieved at scale. Global ore grade decline increases the energy intensity of extraction and processing. Innovation can help make mines less energy-intensive and less environmentally damaging, but innovation can slow the permitting process, since regulators may be unfamiliar with new technology and hesitant to approve permits based on technology that has not been proven out.[8]

Developing robust supply chains will require more than technical solutions along the supply chain. These projects require large up-front investment and have long lead times, and price volatility creates risk, deterring private investors. Traditional financing often struggles with these uncertainties, as critical mineral markets are small, opaque, and subject to price manipulation. Fragmented and sparse Western supply chains expect each segment (i.e., miners, refiners, component makers) to be independently profitable, introducing higher costs and vulnerability (figure 1.2).

Over the decades of offshoring, the United States has also lost expertise and human capital in mining, metallurgy, and manufacturing. There is a shortage of college- and graduate-trained metallurgists,

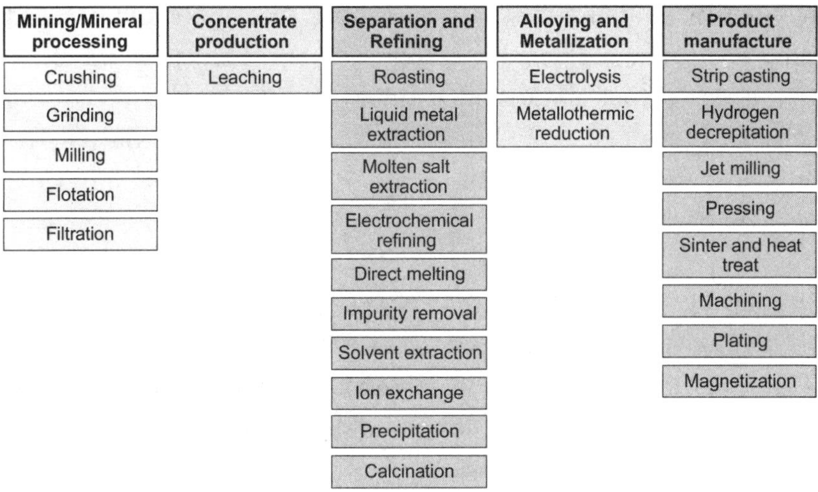

Figure 1.2
The complete supply chain from raw material extraction to end-product manufacturing. Each block represents a major process step, such as mining, processing, refining, and manufacturing, with its associated sub-steps grouped beneath to highlight their connection within the value chain.

mineral processing engineers, and technicians in the critical minerals space. Additionally, specialized equipment for mineral processing, refining (e.g., chemical reactors, separation equipment), and end-product manufacturing (i.e., rare earth magnet manufacture) must be imported. In some cases, necessary equipment trade has been restricted, meaning that reliance on foreign equipment vendors adds another layer of dependency.

Mining and processing have significant environmental impacts (e.g., water usage, tailings generation, mine drainage, and water contamination) and social impacts (e.g., local community effects, labor conditions). Without innovating and aligning societal expectations on what it requires to build manufacturing capability, bringing these industries onshore is prohibitive.

Recommendations

Increasing US competitiveness in critical minerals and materials requires integrated policy action on both the supply and demand side. The timing of policy measures will also need to be carefully calibrated so that a sufficient amount of mining and processing capacity is available domestically before imports are reduced. The following five recommendations speak to each stage of industrial development.

1. **Improve programs in research, development, and demonstration (RD&D) on mineral and metal extraction and processing to increase the domestic supply of critical minerals while reducing their environmental impacts and boosting job training and capacity building.**

Efforts by the federal government to support RD&D for minerals and metals extraction can be made more effective by the following initiatives:

- Engage in technology road mapping across agencies that focuses on the cutting edge of the fundamental science and engineering needed for automated, in situ mining.
- Create well-resourced and digitally coordinated testing capabilities, particularly with the capability to test sustainable in situ mining capabilities.
- Establish comprehensive, centralized data platforms on projects, testing results, and local environmental conditions.
- Build human capital in mining, metallurgy, and manufacturing, which impacts both having enough skilled professionals to design, build, and operate them and lack of innovation because the pipeline of researchers and engineers is thin.

Cross-agency and cross-governmental collaboration on critical minerals is gaining momentum but remains in the early stages of maturity. Internationally, cross-government efforts like the Minerals Security Partnership (MSP), which includes Australia, Canada, Estonia, Finland, France, Germany, India, Italy, Japan, Norway, the Republic of Korea, Sweden, the United Kingdom, the United States, and the European Commission,[9] have focused on accelerating the development of resilient and diverse global critical mineral supply chains. For example, MSP developed the MSP Forum to support individual projects that align with their strategic goals, but current efforts remain largely advisory, lacking enforcement power or executive authority. To stabilize supply chains and move from strategic intent to action, the United States could benefit from establishing a centralized mineral authority with executive authority. A more centralized approach could enable the kind of international coordination necessary to avoid outcomes like the MSP-backed rare earth project in Brazil, which was ultimately compromised by prior Chinese offtake commitments. Models like the Japan Organization for Metals and Energy Security (JOGMEC),[10] where state-backed financing is paired with guaranteed offtake, offer examples of a strategic, government-led approach for long-term resource security.

Within the United States, the Department of Energy has seventeen offices that intersect with critical minerals efforts and another sixty across the government. One promising initiative is the Critical Minerals Collaborative, a coordination mechanism launched in 2023 that brings together Department of Energy offices, other federal agencies, and research stakeholders to align critical materials RD&D programs across the innovation ecosystem.[11] This nascent but effective effort should be further resourced and deepened. To enhance its impact, the cross-agency collaborations should yield outcomes such as shared datasets that are required to demonstrate interoperability, software platforms that reduce barriers to collaboration, terminology that is understandable by all parties, and metrics that quantify the benefit of coordination. With these critical foundations, a coordinated technology road map can be outlined to align R&D and commercialization priorities. In parallel, the government could offer centralized databases for existing capital and machinery to test processes and build capacity. The government can also encourage cross-industry learning by bringing together those with subsurface expertise in adjacent industries, such as fossil fuels, to accelerate technology transfer and workforce development.

Regional collaborative funding efforts provide a basis for technology road mapping as well as ways to field-test technologies through test sites specifically tailored for advanced technologies. Some examples include the Carbon Ore, Rare Earth, and Critical Minerals Initiative,[12] which focused on deriving valuable products from coal, or the Carbon Storage Assurance Facility Enterprise,[13] which sought opportunities in carbon dioxide storage. Sites for testing new technologies should be required to use data structures and software that can enable accessible, clean, and interoperable datasets and data standards to build digital platforms for technology innovation. Examples could be drawn from the Battery Materials Passport efforts emerging in the European Union or domestic efforts such as the Earth Mapping Resources Initiative.[14] The government could also hire advanced analytics teams to help mining experts interpret data in a way that will make it useful in their own contexts.

University programs in mining and metallurgy have either shrunk or been closed, and China has been actively developing expertise. Without addressing the workforce issue, other efforts stall. Programs are needed to build talent pools focused on curricula that align with emerging needs and tie workforce development to actual projects through partnering with proximate colleges to train technicians. Programs could also increase know-how by tapping into the knowledge of retired experts through mentorship programs or rotational programs. Ultimately, in order to increase enrollment in relevant programs, we must raise the profile of these careers through competitive salaries and the promotion of skills in mining and metallurgy as strategically important to the US economy.

2. Implement demand-side strategies that encourage sustained manufacturing scale-up.

Policies that build demand for domestic critical materials can provide signals to mobilize and strengthen domestic supply chains. For example, the investments made under the Inflation Reduction Act have significantly increased investment in US mining, processing, and battery cell manufacturing,[15] but new investments are needed, and current investments must be sustained. The federal government should pursue the following pathways for sustained investment: federal procurement, strategic reserves, demand-side technology road mapping, and reconfiguring demand-side policies.

Federal procurement can stimulate demand, protect new manufacturing from fluctuating global prices, and create a sort of strategic reserve of minerals if the government purchases minerals and precursor materials when prices are low. These materials can then be made available later when prices are higher. Purchases can be targeted using analytics to inform the likelihood of longer-term volatility in minerals of interest. Advance market commitments can also ensure stability in demand for domestic mining. These could manifest as funding of demonstration projects to build confidence in new technologies needed to extract and process more efficiently and build competitive projects. These demonstrations should be based on cross-agency coordinated technology road mapping (similar to coordinated efforts on critical minerals supply) to leverage research support but avoid redundancy. Federal support currently focuses on larger companies that are not necessarily US-owned (even though they are operating in the United States), which means there are missed opportunities to capitalize longer-term manufacturing capability where there has been R&D innovation (often based on federal investment) in advanced technologies such as advanced nickel chemistries and solid-state batteries.

A strategic reserve for critical minerals could play a role in insulating the US economy from future supply shocks—an idea that has gained bipartisan attention in recent years. This is like the Strategic Petroleum Reserve developed in the 1970s, which was established to address shortages and has more recently been used to respond to geopolitical disruptions and natural disasters. Similar guidelines could be created for modern supply chain vulnerabilities, particularly for critical minerals necessary for energy, defense, and technology.[16] More recently, discussions have expanded to include the idea of a US sovereign wealth fund as a strategic investment vehicle.[17] While states like Alaska, Texas, and New Mexico have their own funds for financing their education systems and state government operations, no equivalent exists at the federal level. A national fund could provide targeted investments to stabilize critical supply chains and protect against future disruptions.

Policies that provide preferential procurement to drive demand should encourage innovation and capacity building not only at the manufacturing and processing stages of materials development, but also across the supply chain, including device engineering, such as battery pack design. This increased vertical capability would enable high-performing chemistries to benefit from the same manufacturing learning and engineering capabilities that are currently giving chemistries such as lithium iron phosphate a growing market in China.

Additionally, some forms of materials or minerals could be substituted in battery chemistries in ways that strengthen US supply chains. For example, increasing the proportion of synthetic graphite in anodes can help reduce reliance on natural graphite, which is primarily produced in China (while synthetic graphite could be produced in the United States). Additionally, if graphite demand is made flexible by enabling the use of lower-quality graphite on the anode, it could reduce some of the processing requirements and costs. Reduced processing requirements and costs can ultimately support a sustained manufacturing industry, and the federal government and innovation are part of the path forward.

3. Encourage investment in the recycling of products that contain critical metals.

The United States is a net consumer of critical materials, which should create a rationale and a market for recycling. But to take advantage of end-of-life materials, greater investment is needed in all stages of the recycling industry—collecting devices from products, disassembling those devices, harvesting constituents from those devices, and processing each constituent into marketable materials or components for reuse in new components or products.

A comprehensive strategy will need to account for feedback effects—as supply increases, prices are reduced, making recycling less profitable at least until economies of scale in recycling develop. Sustained investment will be needed while capacity is being constructed, and collection infrastructure responds to market dynamics. Further, as battery chemistry for certain applications shifts to lower-value cathode chemistries, such as lithium iron phosphate, investment will be needed in innovative recycling approaches and cost-effective collection. Even with such a shift, lithium recovery could still provide economic value, as could processes that directly lithiate (i.e., chemically integrate lithium into the structure of the cathode) so that the metal value would not have to be the sole aspect contributing to the economic feasibility of recycling technology. Particular attention is needed to the recycling of graphite. Currently, China is almost the exclusive source of the material, and cost-effective recycling processes for graphite do not yet exist.

The availability of end-of-life materials potentially available for recycling can be predicted based on what is being produced. For example, if it is known how many electric vehicles (EVs) are being sold and what their average lifetimes are, it is possible to estimate how many EVs will reach their end of life in ten years. (Other end-use opportunities may develop as electrification extends to other modes of transportation, including aviation and shipping.) We can therefore build infrastructure now for the feedstocks that will be coming. Given that recycling capability must be built out now for products that will be reaching their end of life in the future, public/private partnerships should encourage the creation of recycling capacity for materials that are now in use, and the potential of recycling should be considered when creating policies for new mining.

In the meantime, requiring that scrap from battery manufacturing be processed domestically could provide feedstock materials to battery recycling facilities. Scrap rates from battery manufacturing (as opposed to end of product life) can be upward of 50 percent in the first one to two years of battery plant operation, although they eventually drop to much lower numbers. Battery scrap from manufacturing is not a sustainable long-term source of recyclable materials but can help bridge the gap as recycling facilities scale up.

Recycled lithium-ion batteries and recycled copper are two areas that offer significant potential for increasing the supplies of these two critical metals. For lithium-ion batteries, the United States should create regional systems for the collection and processing of materials for recycling. It should also promote standard and transparently communicated pack design, battery management systems, and labeling. Rules should require standardized labeling and descriptions of battery form factors (e.g., cylindrical, prism, and pouch) and chemistries so that recyclers can easily tell what approaches to use for recovery and processing. That would not limit innovation in battery chemistry and pack design. The government should require that recycling-critical information be made visible on products (such as quantity of certain elements, or regularity within form factors to increase automated processing). The recycling system should be regional to save money by reducing transportation distances and to streamline the collection process. Similar initiatives are already taking shape internationally, particularly in Europe.[18] The European Commission adopted a new EU Battery Regulation (August 2023), which introduces a digital battery passport with information on capacity, performance, durability, and chemical composition.[19] It also mandates disclosure of recycled content, including a minimum threshold of 6 percent recycled lithium. At the global level, the Basel Convention was updated in August 2024 (signed by 191 countries, though not ratified by the United States) to call for the development of an inventory of waste batteries containing lithium.[20] Meanwhile, the Global Battery Alliance is advancing its own version of a battery passport to improve traceability and sustainability in battery value chains,[21] although the United States is not yet a participant in that effort.

For copper, there is already a significant supply of copper recovered after use (from such sources as computer and communications products), but it is not utilized. The United States does not have the smelting and refining capacity to turn this scrap copper into newly usable material for potential customers such as semiconductor manufacturers. Our estimates indicate that recycled copper could significantly help reduce demand for new copper. By 2035, we expect there to be approximately 1.1 million metric tons of cumulative uncollected copper secondary material in the United States because of increased demand for energy and consumer electronics, as well as material from the building and construction sector. Copper from printed circuit boards will be an important source of scrap copper. Globally, we have found that there is an opportunity to increase scrap supply in 2040 by 44 percent (6.0 metric tons) above current levels, which would help reduce demand for newly mined copper significantly.

In addition, the United States is likely to have more scrap copper available because it is less likely to be shipped abroad as a result of increased import and export restrictions and increased domestic scrap generation in China. Demand will also be more certain because there has been only lukewarm interest in investment in mining copper, while demand for copper is expected to grow. For example, the electronics and automotive industries have made commitments to use more recycled copper in their products and could realize those commitments domestically if copper refining infrastructure began to ramp up now.

The federal government should participate in public–private partnerships to sustain engagement in copper refining in the United States. Companies such as Aurubis in Georgia and Wielend in Kentucky are creating domestic recycling capability and manufacturing semifinished products for customers throughout North America.

4. Implement standards and best practices for permitting mines and early-stage community engagement in the process.

Permitting mines in the United States takes seven to ten years on average, compared with two years on average in Canada and Australia.[22] Review times could be shortened by requiring federal permitting agencies to provide more certainty and transparency on review times, while providing the agencies with the resources and infrastructure necessary to conduct timely reviews.

The relevant federal agency or agencies could be required to negotiate a timeline with mining companies at the outset of a project, along with enforcement mechanisms if deadlines are missed. In addition, state permitting agencies would benefit from more federal assistance to improve timelines.

Making permit review timelines publicly accessible is also helpful. The Permitting Dashboard is one example of regulatory infrastructure that improves transparency, but it remains underused by the mining industry.[23] The Fixing America's Surface Transportation (FAST) Act established the Federal Permitting Improvement Steering Council in 2015, with Title 41 creating a transparent permitting process for covered infrastructure projects. However, FAST-41 has only recently begun to be used as a policy mechanism for streamlining mine permitting. The National Energy Dominance Council submitted twenty projects in the spring of 2025.[24] These include exploration projects, such as McDermitt (lithium) in Nevada, open-pit mining operations such as NorthMet (nickel–copper) in Minnesota, and operational expansions such as the Silver Peak lithium mine in Nevada.

The federal government should also develop frameworks for more rapidly and methodically evaluating new mining technologies. For example, technology scorecards could be developed to compare new technologies with existing ones rather than requiring an entirely new review. Now, perversely, innovation introduces uncertainty and delay if regulators are unfamiliar with a new technology and its impacts on local communities and environments. Increased simulation and modeling capabilities could also inform the permitting of new technologies.

Engaging local communities early and meaningfully in the development process is critical not only for building trust and aligning projects with local priorities, but also for improving permitting timelines and reducing the risk of future delays or opposition. Without meaningful engagement, projects may face costly setbacks, legal challenges, or long-term reputational harm that can derail development. One promising approach is the use of Good Neighbor Agreements (GNAs),[25] which empower community groups with decision-making authority and create adaptable frameworks to accommodate mine decisions over the lifetime of the operation. The mine permitting process allows mines to change over time with very few permit-issuing decision points. A GNA allows community groups to be involved in that iterative process. GNAs do impose higher up-front costs but can ultimately streamline the permitting process.

5. Require that mineral sources be traceable and the supply chain transparent and require that minerals used in the United States meet global industry standards.

Environmental safeguards and enforcement mechanisms within the United States could help ensure that critical materials mining and production are more environmentally responsible, which could be turned into a competitive advantage through innovative reduction of wastes, participation in standards efforts, and providing access to premium markets. But currently, this is hampered by the lack of visibility into the supply chain, which erodes interest by demand-side actors who seek commodities at the lowest price while externalizing environmental impacts.

The current voluntary standards create little to no market pressure for sustainable production, leading to uncompetitive domestic mining and processing capabilities. Government and/or industry requirements are needed to use traceability models, along with global standards for discerning the social and environmental attributes of the minerals supply.

Such approaches have proven successful. For example, US disclosure regulations for conflict minerals led to responsible sourcing standards and certification schemes.[26] In response, major consuming industries, particularly electronics, modified their supply chains, creating demand for conflict-free minerals. This approach depends on the development of an effective regulatory infrastructure.

Industry standards for responsible mining are being developed by international organizations and metal-specific multistakeholder initiatives and institutions, and these organizations conduct voluntary audits of mine sites. However, few projects have undergone audits because participation is voluntary, and underlying conditions differ among regions. As a result, the impact of these initiatives has been quite limited.

The United States could require that imported critical minerals and the products that use them comply with international standards. In addition to its other benefits, this would create a more level playing field for domestic mineral producers, which must meet higher standards. The United States should take an active role in global standard-setting efforts across the supply chain in areas such as product standards and shaping international norms for a wide range of critical minerals issues, including product quality and testing procedures, supply chain governance, traceability, transparency, and sustainability.[27]

Conclusion

The United States is at a strategic crossroads in its effort to build secure and resilient critical mineral supply chains. Both industry and government must act decisively and collaboratively to close the substantial gaps identified in this chapter. For the government, this means establishing coordinated, well-resourced frameworks that connect permitting reform, technology investment, workforce development, international diplomacy, and demand-side policies. Targeted public investments in recycling infrastructure, early-stage technology demonstrations, and mineral reserves, paired with transparent and enforceable environmental and sourcing standards, can provide the stability and long-term direction the private sector needs to de-risk engagement and scale domestic capacity. Additionally, the federal government must coordinate across its many agencies, streamline permitting processes, and expand collaboration with allies through mechanisms that include data sharing, coordinated offtake strategies, and investment alignment.

At the same time, industry needs to move beyond short-term cost considerations and take proactive steps to invest in domestic refining, recycling, and manufacturing capabilities that align with long-term national interests. The private sector can lead in adopting traceability tools, participating in standards development, and piloting innovative technologies with the support of the public sector to address uncertain regulatory and market environments. The private sector can also lead by supporting a robust talent pipeline through partnerships with universities and community colleges, helping to rebuild the technical workforce in mining, metallurgy, and advanced manufacturing. Companies that prioritize responsible sourcing, efficient product design, and collaborative permitting practices will be better positioned to meet future regulatory requirements and compete in premium markets.

Together, government and industry have the opportunity to shape a resilient future where the United States not only secures its own critical mineral requirements but also sets the global standard for twenty-first-century responsible mineral stewardship. Achieving this vision will require strategic alignment, bold investments, and a shared commitment to advancing innovation, infrastructure, and incentives across the entire value chain.

The author would like to thank the following MIT colleagues for their contribution to this chapter: Karan Bhuwalka, former PhD researcher, Mechanical Engineering; Elizabeth Moore, Program Manager, Department of Materials Science and Engineering; Abigail Randall, former master's student, Technology & Policy Program; and Basuhi Ravi, former PhD researcher, Materials Science & Engineering.

Figure 1.3
Global supply of key critical minerals by country in 2024, showing both mining and refining capacity in kilotonnes. (DRC = Democratic Republic of Congo). Data from International Energy Agency, Global Critical Minerals Outlook 2025 (IEA, 2025), https://www.iea.org/reports/global-critical-minerals-outlook-2025.

Appendix

Copper
Mining
Refining
China
Chile
DRC
Peru
Russia
United States
Japan
India
Rest of world
8,000
3,000
2,000
7,000
12,000

Lithium
Mining
Chemicals
China
Chile
Australia
Argentina
Zimbabwe
Rest of world
100
50
0
50
100
150
200

Graphite
Mining
Battery grade
China
Mozambique
Madagascar
Brazil
Russia
Tanzania
Japan
United States
Rest of world
2,000
1,500
1,000
500
0
500
1,000
1,500
2,000

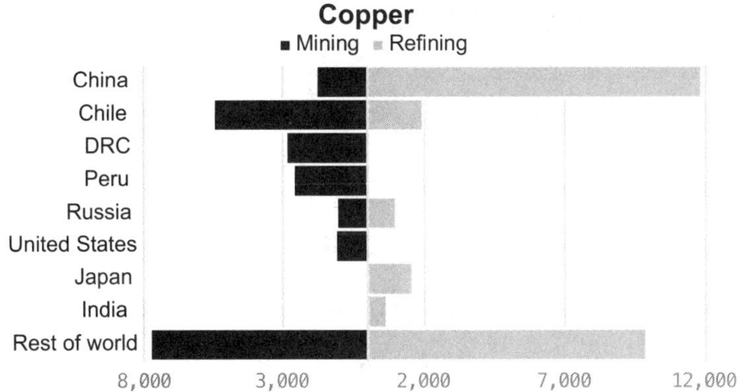

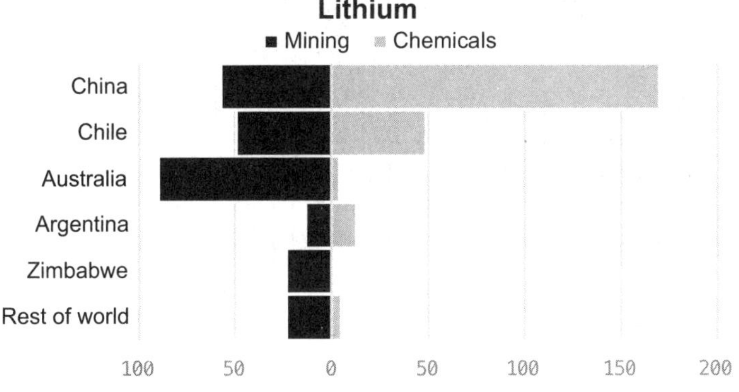

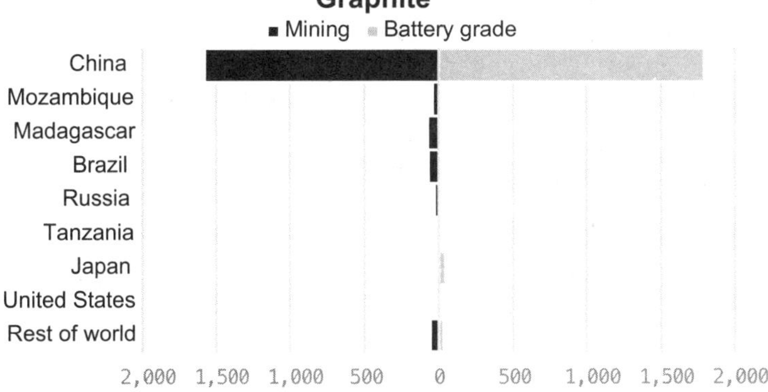

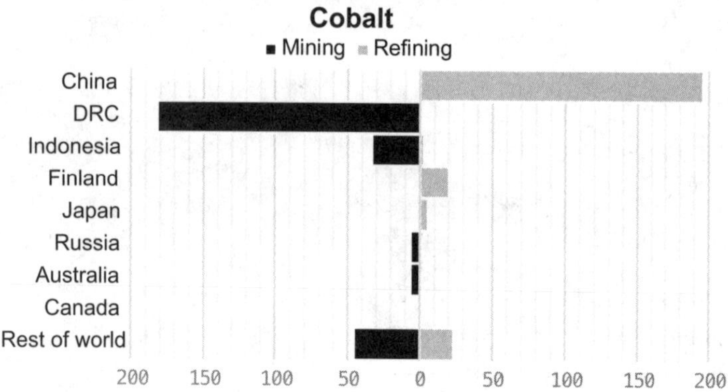

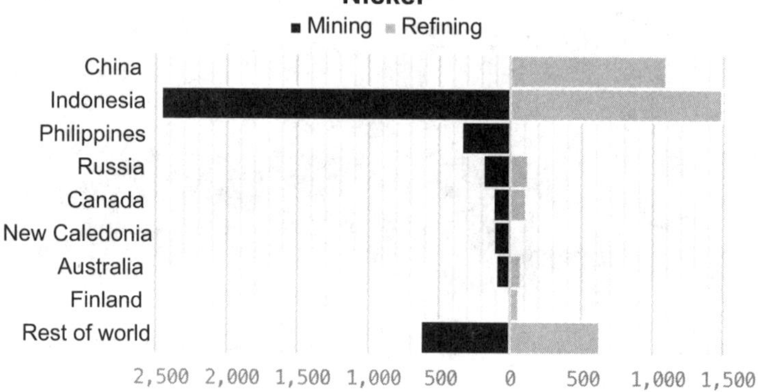

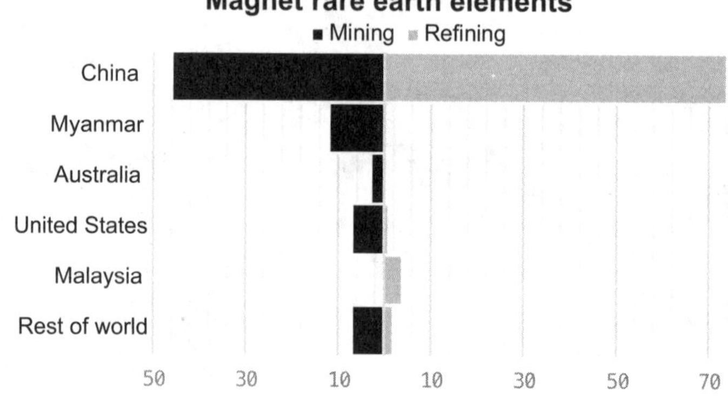

Figure 1.3 (continued)

Notes

1. International Energy Agency, Global Critical Minerals Outlook 2025 (Paris: IEA, June 2025), https://iea.blob.core.windows.net/assets/ef5e9b70-3374-4caa-ba9d-19c72253bfc4/GlobalCriticalMineralsOutlook2025.pdf.

2. Select Committee on the Strategic Competition Between the US and the Chinese Communist Party, "Reset, Prevent, Build: A Strategy to Win America's Economic Competition with the Chinese Communist Party," December 12, 2023, 3, https://docs.house.gov/meetings/ZS/ZS00/20231212/116682/HRPT-118-3.pdf.

3. US Geological Survey, "U.S. Geological Survey Releases 2022 List of Critical Minerals," February 22, 2022, https://www.usgs.gov/news/national-news-release/us-geological-survey-releases-2022-list-critical-minerals.

4. US Geological Survey, *Mineral Commodity Summaries 2024* (US Geological Survey, 2024), https://doi.org/10.3133/mcs2024.

5. US Geological Survey, "U.S. Geological Survey Releases 2022 List of Critical Minerals," February 22, 2022, https://www.usgs.gov/news/national-news-release/us-geological-survey-releases-2022-list-critical-minerals.

6. Gracelin Baskaran and Meredith Schwartz, "The Consequences of China's New Rare Earths Export Restrictions," Center for Strategic and International Studies, April 14, 2025, https://www.csis.org/analysis/consequences-chinas-new-rare-earths-export-restrictions.

7. For a map showing focus areas for twenty-three mineral systems that could host critical mineral resources in the United States and Puerto Rico, see Jane M. Hammarstrom, Douglas C. Kreiner, Connie L. Dicken, and Laurel G. Woodruff, *National Map of Focus Areas for Potential Critical Mineral Resources in the United States*, No. 2023–3007 (US Geological Survey, 2023, https://doi.org/10.3133/fs20233007).

8. Calvo Guiomar, Gavin Mudd, Alicia Valero, and Antonio Valero, "Decreasing Ore Grades in Global Metallic Mining: A Theoretical Issue or a Global Reality?," *Resources* 5, no. 4 (2016): 36, https://doi.org/10.3390/resources5040036.

9. "Minerals Security Partnership," US Department of State, https://www.state.gov/minerals-security-partnership.

10. Japan Organization for Metals and Energy Security, https://www.jogmec.go.jp/english/.

11. "Critical Materials Collaborative," US Department of Energy, https://www.energy.gov/cmm/critical-materials-collaborative.

12. National Energy Technology Laboratory, "The Carbon Ore, Rare Earth, and Critical Minerals (CORE-CM) Initiative," accessed July 25, 2025, https://netl.doe.gov/resource-sustainability/critical-minerals-and-materials/core-cm.

13. US Department of Energy, Office of Fossil Energy and Carbon Management, "Funding Notice: Carbon Storage Assurance Facility Enterprise (CarbonSAFE): Phase II—Storage Complex Feasibility," accessed July 25, 2025, https://www.energy.gov/fecm/funding-notice-carbon-storage-assurance-facility-enterprise-carbonsafe-phase-ii-storage.

14. US Geological Survey, "Earth Mapping Resources Initiative (Earth MRI)," accessed July 25, 2025, https://www.usgs.gov/earth-mapping-resources-initiative-earth-mri.

15. "Clean Investment Monitor," Rhodium Group and MIT's Center for Energy and Environmental Policy Research, accessed July 25, 2025, https://www.cleaninvestmentmonitor.org/.

16. Daleep Singh and Arnab Datta, "Reimagining the SPR," *Financial Times*, February 24, 2024, https://www.ft.com/content/e948ae78-cfec-43c0-ad5e-2ff59d1555e9.

17. Kevin Brennan, Gautam Jain, and Sagatom Saha, "Structuring a US Sovereign Wealth Fund," Center on Global Energy Policy at Columbia University, School of International and Public Affairs, May 1, 2025, https://www.energypolicy.columbia.edu/structuring-a-us-sovereign-wealth-fund/.

18. Maximilian Stephan, "Battery Recycling in Europe Continues to Pick Up Speed: Recycling Capacities of Lithium-Ion Batteries in Europe," Fraunhofer Institute for Systems and Innovation Research ISI, August 7, 2024, https://www.isi.fraunhofer.de/en/blog/themen/batterie-update/lithium-ionen-batterie-recycling-europa-kapazitaeten-update-2024.html.

19. Clare Connellan, Genevra Forwood, Christoph Arhold, Clarissa Marazzi, and William De Catelle, "New EU Batteries Regulation: Introducing Enhanced Sustainability, Recycling and Safety Requirements," White & Case, August 2, 2023, https://www.whitecase.com/insight-alert/new-eu-batteries-regulation-introducing-enhanced-sustainability-recycling-and-safety.

20. "Guidance on Inventory of Hazardous Wastes," Basel Convention, https://www.basel.int/Countries/NationalReporting/Guidanceoninventoryofhazardouswastes/tabid/8755/Default.aspx.

21. Global Battery Alliance, "Battery Passport," https://www.globalbattery.org/battery-passport/.

22. SNL Metals & Mining, "Permitting, Economic Value and Mining in the United States," 2015, https://nma.org/wp-content/uploads/2016/09/SNL_Permitting_Delay_Report-Online.pdf.

23. "Federal Infrastructure Projects Permitting Dashboard," https://www.permits.performance.gov/.

24. The White House, "Trump Administration Advances First Wave of Critical Mineral Production Projects," April 18, 2025, https://www.whitehouse.gov/articles/2025/04/trump-administration-advances-first-wave-of-critical-mineral-production-projects/.

25. Sanford Lewis and Diane Henkels, "Good Neighbor Agreements: A Tool for Environmental and Social Justice," *Social Justice* 23, no. 4 (1996): 134–151, https://www.jstor.org/stable/29766980.

26. Conflict minerals refer to tin, tantalum, tungsten, and gold (3TG), which are tied to armed conflict and human rights abuses in certain sourcing regions.

27. Craig Hart, "Mapping China's Strategy for Rare Earths Dominance," Atlantic Council Global Energy Center, June 13, 2025, https://www.atlanticcouncil.org/wp-content/uploads/2025/06/Mapping-Chinas-strategy-for-rare-earths-dominance.pdf.

2
Semiconductors: Rebuilding the US Semiconductor Ecosystem for Sustained Global Leadership

Jesús A. del Alamo

Strategic Importance

Semiconductors are the oxygen of modern human society. They permeate virtually all areas of human activity: health, transportation, energy, communications, education, defense, entertainment, manufacturing, and many more. As with oxygen, we are not aware of how essential semiconductors are to contemporary life until we don't have them. We gained an appreciation of this through the supply disruptions that took place in 2021 after a fire in a Japanese memory chip manufacturing plant (known in the industry as a fab—short for fabrication plant)[1] and then the loss of power to several fabs due to ice storms in Texas.[2]

The global supply chain disruptions caused by the COVID-19 pandemic crystallized in the public mind the pervasiveness and criticality of chips to human society. The poster child of the ensuing debacle was the sight of 60,000 F-150 trucks (the most popular vehicle in the United States at the time) parked at an airport strip in the Midwest. They were nearly fully finished but were unsellable due to some missing chips.[3] In fact, in 2021, the car industry in the United States lost $210 billion in sales due to the lack of chips, while worldwide and across industries, unrealized sales worth $500 billion were reported.[4] Approximately a third of the inflation in the United States in 2021 can be attributed to the shortage.[5] The auto industry provides an excellent example of the pervasiveness of chips. A modern car contains anywhere from 500 to 1,500 chips, with electronics adding up to about 40 percent of the cost of a car in 2020.[6]

Semiconductors constituted an industry of $628 billion per year worldwide in 2024,[7] and the industry is projected to grow to $1 trillion by 2030.[8] In terms of the end customer, in 2022, the communications sector consumed the major share (30 percent) of semiconductors, closely followed by the computer industry (including mainframes and personal computers, at 26 percent). Automotive, consumer, and industrial applications followed (at 14 percent each) and, finally, government (at 2 percent).[9] The broad diversity of applications suggests an equally rich breadth of technologies. The limelight is often stolen by the most advanced leading-edge logic and memory technologies that are used in supercomputers, AI processors, cloud data centers, and the latest cell phones and mobile devices. However, also hugely relevant are "legacy" logic technologies widely used in automotive and industrial applications, as well as many specialized technologies present in communications, defense, and energy management.

Behind this large and diverse chip-making industry is a complex supply chain that spans the entire globe: semiconductor wafers and other materials from Asia; chip manufacturing equipment from the United States, Japan, and Europe; fabs located most often in Asia but also in Europe and the United States; chip packaging in Asia; and chip design centers in the United States and Europe. The resilience of this supply chain was tested like never before during the pandemic as various bottlenecks and single points of failure emerged. As we discuss in this chapter, 100 percent resilience and control of the entire supply chain by a country or even by a group of like-minded allies is unfeasible. However, due to the relevance of semiconductors in economic and military security, the United States has identified a few strategic technologies to be sited on US soil and has launched a broad program—the CHIPS and Science Act 2022 (Creating Helpful Incentives to Produce Semiconductors for America)—to incentivize industry to invest in semiconductor manufacturing in the country.[10]

Of highest priority for the United States is leading-edge logic technology. Its strategic significance emerges from its use in high-performance computing and AI processors that enable weather forecasting, drug discovery, aerodynamics simulations, and the training of the most advanced AI algorithms, among many other uses. Logic technology is based on complementary metal-oxide semiconductors (CMOS)—a reference to the transistor structures that are used. Logic CMOS technology generations are generally catalogued through a "node" number that shrinks as newer and more advanced technology is released. The node number is given the unit of nanometers (nm), as in 7, 5, or 3 nm. The leading-edge CMOS technology available in the market today is the 3 nm node. Progress is relentless. The three leading companies—TSMC (Taiwan Semiconductor Manufacturing Company), Samsung, and Intel—have announced new fabs for 2, 1.4, and 1 nm chip production in the next few years.[11]

Today's most advanced chip manufacturing plant is TSMC's Fab 18, located in Tainan, Taiwan. Fab 18 started production in 2023 and represents an investment of $19.5 billion. It implements TSMC's 3 nm technology. Intel recently announced production start of 3 nm technology in its Oregon and Ireland fabs. Samsung began initial production of 3 nm chips in 2022 but has been plagued with manufacturing yield problems.

The next generation of technology will be the 2 nm node.[12] It will enable higher transistor density and, most importantly, an enhancement in energy efficiency, which is of increasing importance in high-performance computing, AI training, and mobile applications. TSMC is on track to release 2 nm technology to volume manufacturing in two new fabs in Taiwan in the second half of 2025.[13]

Current Landscape

The spectacular emergence of AI large-language models and the supply chain disruptions caused by COVID-19 collided and spawned a perfect storm that forced the United States to confront head-on the erosion of its leadership in chip manufacturing, which had been taking place for quite some time. In fact, by 2022, no new leading-edge logic fabs had been built in the United States for well over a decade.[14] As a result, in the same year, exactly 0 percent of leading-edge logic chips were manufactured on US soil, with 92 percent coming from Taiwan's fabs and the rest from South Korea. Further, the position in the legacy chip market of US chip makers had also been substantially weakened. These are the chips that underpin the auto industry and many other sectors of the economy. All in all, by 2020, while US chip makers held an impressive 47 percent worldwide market share, only 13 percent of the global fab capacity resided on US soil.[15]

In 2022, the CHIPS Act emerged in response to this state of affairs. The rationale behind this explicit example of US industrial policy had several elements. Foremost was the need for secure semiconductors for defense systems. Even though the volume of chips in military hardware is relatively small (less than 1 percent of the world market),[16] they play a critical role in many strategic programs such as the F-35 fighter jet, missile defense systems such as Patriot, drone platforms with AI capabilities, electronic warfare and cyber defense systems, and many others. This breadth of applications involves a diversity of chip technologies based on different semiconductors and processes. Leading-edge logic technology is a primary concern due to its outsized impact on emerging AI-capable platforms. The prospect of virtually the entire world supply of the most advanced semiconductors being produced on an island just 100 miles off the China mainland was too high a risk for US defense leadership to accept.

However, strengthening national security was not the only goal of the CHIPS Act. The COVID-19 pandemic dramatically illustrated the vulnerability of vast swaths of the US economy to chip-supply disruptions. Enhancing economic security by assuring an adequate supply of chips even in the midst of conflict or natural disasters became an additional important consideration in the design and articulation of the CHIPS Act.

In addition, the semiconductor industry creates quality jobs. In 2023, the US semiconductor industry had 345,000 employees on US soil, with each job supporting 5.7 jobs in other parts of the US economy.[17] Semiconductor jobs, relative to other sectors, are also very well paid. The average annual salary in the semiconductor industry in the United States in 2020 was $170,000, while the US median income in the same year was $56,287.[18]

Semiconductor fabs are large and complex operations that require extensive support from suppliers and other companies that offer a panoply of services. As a result, there is a large gravitational pull around chip fabs, which create thriving ecosystems with many opportunities for good jobs demanding a wide range of educational levels.

Semiconductor manufacturing is critical to the US capacity to innovate in this domain. Given the outsized financial investment that goes into a new fab, the specialized tools required, and the R&D behind the manufacturing processes, success depends on rapidly ramping up production and quickly achieving and maintaining high manufacturing yields. This, in turn, requires a continuous level of innovation that must take place right on the manufacturing floor and be able to respond to perturbations in the manufacturing process, address feedback from customers, or take advantage of new, more efficient tools or process conditions. Chip manufacturing thrives on continuous learning and innovation at the factory level, such that, in the words of the legendary CEO of Intel, Andrew Grove, "Without scaling [meaning high-volume manufacturing in the United States], we don't just lose jobs—we lose our hold on new technologies. Losing the ability to scale will ultimately damage our capacity to innovate."[19] For the United States to again lead in semiconductors and sustain that leadership into the future, it must have sizable and advanced manufacturing operations on US soil.

The CHIPS and Science Act, passed with bipartisan support in 2022, outlined several goals, including expanding US production capacity, securing supply chains, strengthening R&D leadership, and expanding the workforce.[20] As of June 2025, $33.7 billion of manufacturing incentives have been awarded to nineteen companies to set up or expand fabs or other facilities in twenty-one states.[21] An additional $1.9 billion worth of proposals are still under consideration. The largest award went to Intel, with $7.8 billion to expand or launch new fabs in New Mexico, Arizona, Oregon, and Ohio (the majority of this grant was converted, with additional funds, into a 10 percent equity stake held by the US government by the Trump administration). Foreign companies are also investing: TSMC will receive $6.6 billion in grants for three new fabs in Arizona, while Samsung has been awarded $4.7 billion in support of a new fab in Texas.

The CHIPS grants leveraged significant private-sector capital. Intel, as a case in point, has committed $103 billion toward the construction of new facilities. Separately, TSMC originally pledged to invest $65 billion in its three new fabs in Arizona; more recently, it has made an additional commitment of $100 billion toward expanding its Arizona site. All in all, over $540 billion in private-sector investment commitments have been pledged—a leverage ratio of approximately 16 private-sector dollars to 1 public dollar. It is notable that some of these investments are taking place in entirely new locales such as New Albany, Ohio, where Intel is set to invest $28 billion to launch two new state-of-the-art fabs. Intel's initial investment is projected to create 3,000 high-tech jobs and an estimated 10,000 additional indirect jobs in the region. It is estimated that 350 Ohio businesses will become part of the supply chain for these plants.

The new fab investments accomplish the most crucial goal of the Act: to build leading-edge chips on US soil for the first time in years. With the current state-of-the-art manufacturing at 3 nm as reference, TSMC is committed to build separate 1.6 nm class fabs at its Arizona site. Intel's new fabs will in turn target 1.4 and 1.8 nm technologies. Samsung's new fab investments in Texas will host 2 and 4 nm processes.

Legacy and mature logic nodes are also attracting investments: Global Foundries is launching a new fab in Malta, New York, to support 12 and 14 nm technologies, while Texas Instruments is building two new facilities in Texas and Utah to deliver 26, 65, and 130 nm products. Memory and specialty technologies of particular importance for defense systems are equally well represented in the portfolio of investments spurred by the CHIPS Act.

The passage of the CHIPS Act launched a sort of arms race around the world in support of local semiconductor manufacturing. In 2024 China announced the launch of the China Integrated Circuit Industry Investment Fund (Phase III) with $47.5 billion in support of developing advanced chips for AI and establishing the entire supply chain within the country.[22] Japan announced investments of $65 billion by 2030 to revitalize its semiconductor industry and in 2022 brokered the creation of the Rapidus Consortium by committing a $4.5 billion investment toward leading-edge logic manufacturing.[23] The European Union in 2023 launched the European Chips Act, endowed with $47.5 billion, with the goal of "doubling the EU's global market share in semiconductors by 2030."[24] South Korea responded with the K-Chips Act, which amounts to $41 billion toward infrastructure funding and talent-development initiatives in support of the Korean industry.[25] Taiwan earmarked $9.3 billion over nine years to foster several areas, including integrating generative AI with silicon chips, cultivating talent, and attracting international startups and investment.[26] Finally, the UK National Semiconductor Strategy is dedicating $1.2 billion over the next decade to strengthen semiconductor design and compound semiconductors, among others.[27]

All in all, the global landscape for semiconductor chip manufacturing will look very different within the next five years, with many countries increasing domestic capacity.

Gaps and Opportunities

The following areas need to continue to receive attention to ensure sustained US leadership in the semiconductor industry.

Research and Development (R&D)

In addition to fostering chip manufacturing on US soil, in an environment of rapidly changing technology, it is imperative that the United States seeks not only to reestablish leadership in semiconductor manufacturing but also to invest in frontier R&D, where it has historically led the world. It is unclear how funding dedicated to R&D in the CHIPS Act will be allocated at this point in time but it is essential for the country to have a national strategy around accelerating US R&D capacity and building out the semiconductor ecosystem, including creating cutting edge facilities for fast prototyping of new concepts and expanding the pipeline of skilled workers.

It will be particularly important to advance US leadership in an emerging technology of great strategic significance: packaging based on chiplet integration technology. This refers to a new modular approach to building electronic systems by integrating separately manufactured "bare" chips into a compact, higher-level assembly known as "system-in-package." Chiplet-level integration has many advantages over previous approaches, such as board-level integration of individually packaged chips or the "system-on-chip" route in which different functions such as memory, logic, and communications are all implemented on the same chip. In contrast with these earlier schemes, chiplet-level integration promises higher system performance, greater energy efficiency, more compact form factor, faster design cycles, and lower costs, as individual chiplets are fabricated on the most suitable cost-optimized technology and, in many cases, already exist and can be taken off the shelf.

Individual chip packaging has long been considered a relatively low-tech, low-profit-margin activity mostly carried out in Southeast Asia. Chiplet-based packaging technology, on the other hand, is far more sophisticated, involving the high-precision assembly of multiple very high-value dies using dense micron-scale 3D chip interconnects. The world leader at the present is TSMC, which, from packaging facilities located in Taiwan, is already supplying chiplet assemblies for advanced AI processors to Nvidia, AMD, Apple, and other companies. In the United States, Intel has developed a similar technology in-house that is already being used in some advanced processors.

Regarding all this, a key goal of the CHIPS Act is to establish a domestic advanced packaging capability to be able to package in the United States advanced-node chips that have been manufactured in the country itself. Developing this capability in the United States will be critical to securing US capabilities across the semiconductor value chain.

Semiconductor Workforce Development

The large expansion of the semiconductor industry in the United States is predicted to create many new jobs; however, after a history of decline, it will be a challenge to fill these jobs. A recent study forecasts a serious labor shortage over the next five years. While the semiconductor industry will create 160,000 new jobs in the United States (88,000 engineers and 75,000 technicians), at current graduation rates, the supply is projected to be much smaller: 11,000 engineers and 7,000 technicians by 2029.[28] The CHIPS Act program and the major chip manufacturers are putting in place many programs to bridge the technician gap. If these are successful, the technician shortage could be averted. Such is not the case for the dearth of engineers, which, even in the most successful scenarios, would leave a very large gap. This is going to be a huge challenge, and a quick solution is not apparent at this point.

It is not just the numbers, but also the composition of the semiconductor workforce that is a concern. Perhaps reflecting the attitude of US chip makers, the then-Secretary of Commerce, Gina Raimondo, stated in 2023 that "[more] than 60 percent of the jobs in a fab don't require a college degree."[29] As it turns out, this is very different from what TSMC, the leading chip maker in the world, actually practices. TSMC data from 2020 shows that only around 19 percent of jobs do not require a college degree and that, in fact, 51 percent of the jobs require graduate degrees. The equivalent figures for the US semiconductor industry as a whole is nearly inverted: 44 percent and 26 percent, respectively.[30] This profound difference in the composition of the semiconductor workforce will necessarily have a huge impact both on manufacturing operations and on the innovation and learning that takes place on the manufacturing shop floor.

Cost of Manufacturing

An enduring challenge for sustained US leadership in semiconductors is the high cost of ownership of fabs in the United Sates. In a joint April 2021 study by the Boston Consulting Group and the Semiconductor Industry Association,[31] the total ten-year cost of ownership of a fab in Asia (other than in China) was 29 percent cheaper than in the United States. Further, China enjoyed an additional 23 percent cost-effectiveness over the rest of Asia. The majority of this cost advantage (70 percent) is attributed to government incentives, with smaller components coming from reduced construction, labor, and utilities costs. It is precisely because of this that 56 percent of the fab capacity of US chip manufacturers was actually located offshore.[32] At the sunset of the CHIPS Act, this disparity in cost structure is likely to remain, and ensuring continued investments in semiconductor manufacturing in the United States in the absence of federal subsidies will once again become problematic.

Energy Security and Sustainability

The chip industry has an additional problem, not specific to the United States but threatening to curtail its expansion on US soil. The problem is sustainability. Chip making consumes considerable amounts of electricity. This can be best appreciated by looking at Apple's total carbon footprint, which in 2019 amounted to 25.1 metric tons of CO_2 equivalent. Among all possible sources of emissions (business operations, total energy consumed by all Apple products in the world, and energy utilization in hardware manufacturing), the largest component, a full 33 percent of all CO_2 emissions, came just from making the chips that entered Apple's products in that year. And things are only getting worse. As semiconductor technology advances, energy consumption of chip making increases rapidly. For example, in a study by IMEC, the amount of energy required to process a 300 mm wafer (the current state of the art) has increased by a factor of three from the 28 nm node (circa 2010) to the 3 nm node (circa 2024).[33] This trend is only likely to worsen as more complex tools and longer manufacturing cycles will be required in future nodes.

United States–China Trade Conflicts

The ongoing United States–China tech competition could also create challenges for the US semiconductor industry. In an effort to limit China's ability to fabricate advanced chips that could be used to train very large AI models and enable strategic military uses, the United States has restricted China's access to leading chip manufacturing equipment. China has responded by curtailing exports of critical minerals essential for the semiconductor industry. How far this will go is unclear. As China continues to pursue access to leading-edge semiconductor manufacturing equipment and the United States tries to prevent this through all manner of restraints, disruptions to the complex microelectronics supply chain can be expected.

US Foundry Capabilities

The CHIPS Act cannot be considered successful if it fails to establish a US-based semiconductor foundry with leading-edge capabilities that is responsive to US national security needs and can compete head-to-head with TSMC, the de facto monopoly today. It is not an exaggeration to see this as an overwhelming strategic national security interest. Even though US Department of Defense (DoD) systems amount to only about 1 percent of the chip market, emerging AI platforms, electronic warfare, crypto, intelligent drones, and other advanced capabilities that are deemed critical in future warfare demand the most advanced chips and require that they are produced in a secure and reliable environment.

Intel Foundry, the chip manufacturing arm of Intel, is perhaps the most plausible company to fill this role. However, the challenges that Intel Foundry faces are monumental.

For years, Intel has followed a manufacturing philosophy that they call "copy exactly." Essentially, the fabrication process and all its details are developed at Intel's Technology Development operation in Oregon. When it is ready, it is copied exactly, down to the most minute details at the various fabs. Any later changes that might need to be introduced have to be performed first in Oregon and then transferred to the manufacturing plants. For many years, this has worked well for Intel. Until very recently, their fabs have had only one customer, Intel itself, and they have fabricated a small product mix—mainly central processing units (CPUs).

TSMC has a different philosophy that is best characterized as "continuous improvement." TSMC's fabs are empowered to constantly fine-tune the manufacturing process in response to customers' needs and also to adapt to the availability of new tools that might increase manufacturing efficiency. This approach is dictated by the fact that they have a large product mix from multiple customers, with both changing all the time. This manufacturing philosophy—which requires sophisticated technical decision-making rather than just precise replication—explains why TSMC's fabs employ a sizable number of engineers and scientists, many with advanced degrees. Continuous improvement in manufacturing requires continuous innovation at the factory floor, as Andrew Grove emphasized.

The traditional Intel approach of "copy exactly" is not capable of responding in an agile way to problems and opportunities that might arise on the manufacturing floor. Any production process changes must be first validated at the mother ship and then pushed out to the various factories. This is too slow for a foundry business. This dissonance between Intel's traditional manufacturing philosophy and the needs of a foundry business have been recently recognized by the Intel Foundry's new leadership.[34]

Another important factor as Intel Foundry attempts to challenge TSMC's preeminence as a semiconductor foundry is the much greater volume of wafers that TSMC manufactures; for example, in December 2020, Intel produced 0.9 million wafers versus TSMC's production of 2.8 million in the same month. The gap today must surely be much larger. This discrepancy means that TSMC is likely traversing the learning curve at a much faster rate than Intel. Over the years, this adds up to a huge advantage.

All this makes clear the magnitude of the challenge for Intel Foundry to establish itself as a competitive alternative to TSMC.

Nevertheless, in addition to the US government, chip design houses such as Nvidia, Apple, and AMD are hoping that Intel can establish itself as a reliable second source for their products, particularly one that manufactures chips on US soil.

Recommendations

The CHIPS Act, laudable and unprecedented as it is, represents only a stop-gap measure. In a best-case scenario, the policy actions to date will successfully restore a leadership position for the United States in semiconductors. However, in the absence of a thoughtful and deliberate follow-up program, these gains could quickly erode. As semiconductor policy experts and researchers Flamm and Bonvillian assert, "maintaining leadership in semiconductors is a long game,"[35] and the US government must take a long-term interest in nurturing the industry by supporting a thriving ecosystem on US soil. The following are some actions that seem particularly appropriate.

1. Establish and sustain US foundry leadership.

The United States must continue to support US-based world-class chip foundry manufacturing at the leading edge, particularly at a capacity that can serve US national security interests.

- *Provide sustained government support.* The US Department of Defense and other national security agencies should be able to justify sustained financial support of a US-based chip foundry operation that provides reliable and secure access to leading-edge node technology.
- *Foster private-sector consortium support.* Large US companies whose business thrives on high volumes of the most advanced chips, such as Nvidia, Amazon, Apple, and AMD, should have

a strong interest in preserving a competitive second source for their chips on US soil. A consortium to support Intel Foundry through long-term contracts or direct investment could help Intel invest in building leading-edge fabs and in aggressively pursuing R&D in future advanced nodes. Such consortia are not a new concept in the industry. In the early 2010s, Intel, TSMC, and Samsung joined forces to heavily support ASML, a Dutch company, in the development of EUV lithography, *the* crucial technology to advance Moore's Law but a hugely expensive and risky bet.[36] These three chip companies took equity positions in ASML and directly funded R&D on EUV lithography to the tune of about $6 billion.[37] The payoff has been handsome, with EUV enabling continuous node scaling for several generations.

- *Transform manufacturing culture and philosophy.* Intel Foundry must change its internal manufacturing culture from one designed to serve a single customer with a relatively narrow range of products to a new approach that caters to many customers and an ever-changing mix of designs. This will require a more agile and educated workforce on the manufacturing floor that can promptly address problems and customer concerns and take advantage of opportunities brought up by tool upgrades or altogether newer tools. As recently recognized by its new senior leadership, Intel Foundry must speedily transition its manufacturing philosophy from "copy exactly" to "continuous improvement."

2. Invest vigorously in semiconductor R&D.

Ann Kellegher, Intel CTO at the time, stated a while back, "Moore's Law thrives on innovation."[38] It is abundantly clear that while Moore's Law is far from dead, its progress has slowed down. Yet the hunger for ever higher performance computation is going to remain unquenched for a long time. Beyond computing, semiconductor technologies are ubiquitous, and they anchor progress in virtually all aspects of a modern economy. Without doubt, the US competitive advantage in semiconductors resides in its innovation engine, which has been strong in the design phase but less robust in the downstream production stage.

- *Invest in next-generation technologies.* Future progress in the power of electronics will rely less on silicon transistor scaling and more on newer technologies that address bottlenecks in system performance such as 3D integration, chiplet-based system packaging, and integrated photonics. A common thread here is the increasing future role of new devices based on new materials that operate under new physical principles. Microelectronics will continue to bring to market technologies that just a few years back were in nobody's road maps. Sustained investments in fundamental research as well as the physical, digital, and human infrastructure that supports it are needed.
- *Revive and update the "Science" portion of the CHIPS and Science Act.* The "Science" portion of the bipartisan CHIPS and Science Act was approved but funds were never appropriated to secure its goals, which included supporting basic research in semiconductors through multiple R&D programs run by different agencies of the US government. The Science Act should be revived and revised to accommodate the current fast pace of change in the industry, which has accelerated since the Act was first conceived; it also needs updating in light of ongoing CHIPS Act investments.

3. Expand university research and educational infrastructure.
A great deal of innovation in semiconductors has traditionally taken place at US universities and been brought to the marketplace through startups, in many cases founded by the very creators of the new technology. Universities also play a crucial role in training the next generation of scientists and engineers who will eventually propel the industry forward.[39]

- *Strengthen university research capabilities.* Key for universities to continue to contribute to advancing the industry in the United States is the existence of a robust university research and education infrastructure—not only facilities and tools but also the staff support structures that make everything hum. It should then be a national priority to expand and enhance the capabilities of US universities so that they can be the first among their world peers in identifying and demonstrating new concepts, to support the generation of startups that will further develop these technologies and showcase their practical potential, and to train the educated workforce that will bring these new technologies to robust prototyping and eventual volume manufacturing on US soil.
- *Address the shortage of engineers and scientists.* Attracting students to the semiconductor disciplines represents a critical bottleneck in the expansion of the semiconductor workforce essential to foster the industry in the United States. No amount of government subsidies to companies is going to obviate the need for a well-educated and abundant workforce, as industry has well recognized. As noted above, multiple programs under the CHIPS Act are being launched to address the dearth of technicians. No similar attention appears to be given to the dire need for engineers and scientists with bachelor's and graduate degrees.
- *Expand and create new educational and research programs.* Addressing the engineer and scientist gap also calls for expanding existing university programs and creating new ones in institutions that traditionally have stayed on the sidelines of the microelectronics ecosystem. Effective educational programs should rely on hands-on, project-based, design-oriented multidisciplinary research and educational experiences for undergraduates, starting from the first year of college. At the graduate level, research projects should constitute the core of the educational experience. Adequate research funding, a well-equipped and well-run

university research infrastructure, and close connections with industry are the keys to relevance.

- *Leverage world-class US university research as a competitive advantage.* The United States has long served as a magnet that attracts some of the most motivated and best-prepared students from all over the world to pursue advanced degrees in the semiconductor disciplines. Delivering on the immediate and future workforce needs of the US semiconductor industry can only be achieved through a vigorous program to nurture and expand research at US universities that will continue to bring to our shores the most talented from even the farthest corners of the world.

4. Address industry talent acquisition and retention challenges. According to a 2022 report, semiconductor companies face persistent talent issues that threaten to undermine the desire to expand their manufacturing operations in the United States.[40] The report identifies three problem areas: talent acquisition, retention, and organizational health. These issues hamper the recruitment of fresh graduates as well as mid-career shifts from adjoining industries.

- *Improve sectoral brand image.* Semiconductor companies score worse than other technology companies when it comes to brand image and recognition. Students, in particular, perceive that compensation and development opportunities in semiconductor firms are below par. Further, periodic highly publicized layoffs discourage new graduates from considering joining the industry.
- *Address retention and employee satisfaction issues.* Employee loyalty is also a concern, as a sizable fraction of employees express a likelihood of leaving their jobs in the near future. Issues of work–life balance, culture, and values as well as compensation and benefits top the list of employees' concerns.
- *Improve organizational health metrics.* Organizational health, in the report's definition, is the way in which companies and their

employees align with a common vision and execute this vision. In this metric, 64 percent of global companies score higher than semiconductor firms. These also find themselves behind the global benchmark in 34 out of 37 practices relevant to organizational health.

5. Build regional innovation ecosystems.

Sustained US leadership in semiconductors will rely on its rich innovation ecosystem that includes small, medium, and large companies operating at different links of the supply chain, including national labs, DoD labs, nonprofit institutions, community colleges, universities, startups, and venture capital firms. Given the steady consolidation among chip makers into fewer and larger companies, it is the smaller, more innovative, and disruptive companies, developed primarily via university research, that will ultimately transform the industry, often working closely with large strategic partners.

- *Nurture specialized technology foundries.* The key to sustaining future US dominance in microelectronics is the invention and development of "leapfrog" technologies—both those that build on and leverage the current semiconductor paradigm and those that involve disruptive new concepts. Prototyping these on their way to eventual scale-up on US soil requires specialized foundries, also on US soil, with the right mix of equipment, experience, and flexibility. In their absence, new concepts will travel overseas as their developers search for suitable prototyping partners and will likely remain abroad as they transition to volume manufacturing.
- *Exploit regional network efficiencies.* Accomplishing the goals articulated here will involve the engagement of institutions (colleges, universities, community colleges, and small companies) that have traditionally not been part of the US microelectronics enterprise. This can be facilitated by leveraging regional

network effects, that is, pooling existing regional resources and expertise in support of a broader community. Regional networks can create and manage research programs, educational programs, startup support, outreach, and internship programs designed within a limited geographical footprint to facilitate the involvement of educational institutions that were previously on the sidelines. The Microelectronics Commons program of the CHIPS Act is an excellent example of this. It is comprised of a national network of eight regional technology hubs that focus on one or more technology areas defined by the DoD as important to defense microelectronics. Each hub manages a network of commercial innovators in the regional area.

6. Support hardware startups.

Semiconductor startups have had a considerable impact on the development of the industry from its inception. Countless semiconductor startups have flourished into world-class global corporations. Others have been acquired by established companies, which incorporate new technologies at reduced risk and cost. Startups provide extraordinary professional development opportunities for scientists and engineers that not only pave the way for great career paths but also produce positive spillovers for future employers and the industry as a whole. However, hardware startups have greater funding needs and require a longer runway for scale-up than average, which hinders investments by venture capital firms.

- *Incentivize startup creation.* Investment funds need to be created in support of hardware startup formation to de-risk the early development of new technologies and to move ideas more rapidly into higher stages of prototyping and scale-up so as to be able to attract private-sector investment.
- *Subsidize access to university facilities.* Early-phase startup activities could be fostered by partially underwriting access to university facilities, particularly those where the technology was

created by their academic founders in the first place. This allows startups to quickly build on top of demonstrated technologies in their native environment without having to invest large resources in establishing their own facilities only to replicate previous results.

- *Facilitate access to foundries, shared prototyping facilities, and digital design tools.* Incentives are needed to ensure that foundries work not just with larger firms but also with startups. In an industry accustomed to operating at capacity, foundries can pick and choose which startups to work with based on their own strategic calculations. With TSMC, a de facto monopolist, we have a situation where startup innovation in the United States is often gated by the business interests of a foreign corporation. Support from the CHIPS Act offers the US government leverage to negotiate access to leading-edge technologies, if needed, on behalf of US startups. Similar arguments apply to other commercial prototyping facilities and design tool houses.
- *Support an exclusive intellectual property (IP) model for startups.* Exclusive IP is essential to startups for securing the initial capital required to launch and scale companies. The inability to retain IP dissuades the United States' most talented students from attempting to commercialize their innovations upon completion of their studies. Funding models that allow startups and universities to retain their IP rather than pre-competitive consortium models that require shared IP need to be instituted, particularly when government research funds have been used in the research phase.

Conclusion

Semiconductor technology is central to our modern information society. For decades, the United States enjoyed unrivaled leadership in this area. After all, it was invented here. But, over the years, this commanding position has eroded away to the point that a large majority of chips—in particular, the most advanced—are now fabricated overseas. The CHIPS Act represented a concerted effort by the US government to turn back the clock and regain preeminence. Although facing some strong headwinds, current investments will undoubtedly do a lot of good for the country. But we cannot become complacent. Semiconductors are a very long game that thrive in a rich ecosystem, and the technology keeps advancing at a brisk pace. As in the first sixty years, innovation will continue to rule. And as perhaps never before, innovation at the manufacturing floor will play a critical role. The long-term health of the industry demands innovation leadership across the entire ecosystem, and ensuring this will require sustained attention.

The author would like to thank the following colleagues for their contribution to this chapter: Vladimir Bulović, Professor of Electrical Engineering & Computer Science, Founding Director, MIT.nano, and Fariborz Maseeh (1990) Chair in Emerging Technology, MIT; Robert Atkins, former Division Head, Advanced Technology, MIT Lincoln Laboratory; Tomas Palacios, Professor, Department of Electrical Engineering & Computer Science, Director, Microsystems Technology Laboratories, MIT; Sylvia Barmack, former Research Assistant, MIT Shaping the Work of the Future Initiative; Elisabeth B. Reynolds, Professor of the Practice, Department of Urban Studies and Planning, MIT; and Richa Vera Udayana, Master of City Planning student, Department of Urban Studies and Planning, MIT.

Notes

1. Sebastian Moss, "Renesas Semiconductor Fab Catches Fire, Impacting Chip Production," *Data Center Dynamics*, March 22, 2021, https://www.datacenterdynamics.com/en/news/renesas-semiconductor-fab-catches-fire-impacting-chip-production/.

2. Willy Shih, "Severe Winter Weather in Texas Will Impact Many Supply Chains Beyond Chips," *Forbes*, February 19, 2021, https://www.forbes.com/sites/willyshih/2021/02/19/severe-winter-weather-in-texas-will-impact-many-supply-chains-beyond-chips/.

3. Mike Colias, Ben Foldy, and Nora Naughton, "Empty Lots, Angry Customers: Chip Crisis Throws Wrench into Car Business," *Wall Street Journal*, May 13, 2021, https://www.wsj.com/business/autos/empty-lots-angry-customers-chip-crisis-throws-wrench-into-car-business-11620909719.

4. US Department of Commerce, "Analysis for CHIPS Act and BIA Briefing," press release, April 2022, https://www.commerce.gov/news/press-releases/2022/04/analysis-chips-act-and-bia-briefing.

5. US Bureau of Labor Statistics, "Consumer Expenditures in 2021," January 2023, https://www.bls.gov/opub/reports/consumer-expenditures/2021/.

6. See Robert N. Charette, "How Software Is Eating the Car," *IEEE Spectrum*, June 7, 2022, https://spectrum.ieee.org/software-eating-car. Also see Deloitte, "The Key Role of Battery Costs in Automotive: How New Players Are Disrupting the Automotive Industry," October 20, 2023, https://image.marketing.deloitte.de/lib/fe31117075640474771d75/m/1/91247521-57ca-4869-81f5-ad6175d66637.pdf.

7. Semiconductor Industry Association, "Global Semiconductor Sales Increase 19.1% in 2024; Double-Digit Growth Projected in 2025," February 7, 2025, https://www.semiconductors.org/global-semiconductor-sales-increase-19-1-in-2024-double-digit-growth-projected-in-2025.

8. McKinsey & Company, "Semiconductors Have a Big Opportunity—But Barriers to Scale Remain," April 2025, https://www.mckinsey.com/industries/semiconductors/our-insights/semiconductors-have-a-big-opportunity-but-barriers-to-scale-remain.

9. Congressional Research Service, "Semiconductors and the Semiconductor Industry," April 19, 2023, https://www.congress.gov/crs-product/R47508.

10. See "Chips for America," National Institute of Standards and Technology, https://www.nist.gov/chips.

11. An important caveat is relevant at this point (S. K. Moore, "The Node Is Nonsense," *IEEE Spectrum* 57, no. 8 (August 2020): 24–30, https://doi.org/10.1109/MSPEC.2020.9150552). While for several decades the node number roughly coincided with the length of the most critical element of a transistor, its "gate," this correlation long ago stopped applying. Today's 3 nm technology transistors have a gate length of the order 13 nm, and this number is scaling down very slowly from generation to generation. So, as we see industry talking about future 2, 1.4, 1 nm node technologies, there should not be a concern that we are about to run out of atoms! As a result of shrinking dimensions, from generation to generation, transistor density

has doubled. This has enabled ever more capable chips. Today's most advanced chips contain in excess of 100 billion transistors (Venus Kohli, "Final Transistor Count Update of 2024," *PCIM News Platform*, December 18, 2024, https://www.power-and-beyond.com/final-transistor-count-update-of-2024-a-03177edd6bb38cd52259db5457e51025/). Technologies that are in the development stage should allow chips with one trillion transistors around the year 2030 (Mark Liu and H.-S. Philip Wong, "How We'll Reach a 1 Trillion Transistor GPU," *IEEE Spectrum*, March 28, 2024, https://spectrum.ieee.org/trillion-transistor-gpu).

12. This new generation will feature a new transistor architecture, dubbed Gate-All-Around (also other names) that dramatically departs from earlier designs ("planar" down to 32 nm node, "FinFET" down to 3 nm node).

13. Lisa Wang, "TSMC Outlines Work on Nine New Advanced Factories," *Taipei Times*, May 16, 2025, https://www.taipeitimes.com/News/front/archives/2025/05/16/2003836967.

14. Kenneth Flamm and William B. Bonvillian, "Solving America's Chip Manufacturing Crisis," *American Affairs* 9, no. 2 (2025): 41–68, https://americanaffairsjournal.org/2025/05/solving-americas-chip-manufacturing-crisis/.

15. Semiconductor Industry Association, *2021 Factbook*, 2021, https://www.semiconductors.org/wp-content/uploads/2021/05/2021-SIA-Factbook-May-19-FINAL.pdf. Also see Congressional Research Service, "Semiconductors and the Semiconductor Industry," April 19, 2023, https://crsreports.congress.gov/product/pdf/R/R47508.

16. Semiconductor Industry Association, *2024 Factbook*, 2024, https://www.semiconductors.org/wp-content/uploads/2024/05/SIA-2024-Factbook.pdf.

17. Semiconductor Industry Association, "State of the U.S. Semiconductor Industry 2023," 2023, https://www.semiconductors.org/wp-content/uploads/2023/08/SIA_State-of-Industry-Report_2023_Final_080323.pdf.

18. Semiconductor Industry Association and Oxford Economics, "Chipping In: The Positive Impact of the Semiconductor Industry on the American Workforce and How Federal Industry Incentives Will Increase Domestic Jobs," May 2021, https://www.semiconductors.org/wp-content/uploads/2021/05/SIA-Impact_May2021-FINAL-May-19-2021_2.pdf. Income estimates come from Emily A. Shrider et al., *Income and Poverty in the United States: 2020*, US Census Bureau, Current Population Reports, P60-273 (Washington, DC: US Government Publishing Office, September 2021), https://www.census.gov/content/dam/Census/library/publications/2021/demo/p60-273.pdf.

19. Andy Grove, "How America Can Create Jobs," *Bloomberg*, July 1, 2010, https://www.bloomberg.com/news/articles/2010-07-01/andy-grove-how-america-can-create-jobs.

20. The main goals of the Act were (1) to invest in US production of semiconductors, especially leading-edge production; (2) to assure a secure supply of older and current chips for national security and critical manufacturing; (3) to strengthen US semiconductor R&D leadership to capture the next set of critical technologies, applications, and industries; and (4) to grow a diverse semiconductor workforce and strong communities that participate in the prosperity of the industry. The CHIPS portion of the Act contains a $52.7 billion appropriation over five years. The most significant activities are $39 billion semiconductor manufacturing incentives, $11 billion R&D and workforce development programs, and $2 billion to support Department of Defense initiatives. In addition, there are $24 billion in tax credits set aside for investments in semiconductor manufacturing on US soil. See US Department of Commerce, "A Strategy for the CHIPS for America Fund," September 6, 2022, https://www.nist.gov/system/files/documents/2022/09/13/CHIPS-for-America-Strategy%20%28Sept%206%2C%202022%29.pdf.

21. Department of Commerce Office of Inspector General, "Commerce CHIPS Act Programs: Status Report," June 2, 2025, https://www.oversight.gov/sites/default/files/documents/reports/2025-06/OIG-25-021-I%20%28SECURED%29.pdf. Also see US Department of Commerce, "Department of Commerce Announces CHIPS Incentives Award with Hemlock Semiconductor to Help Secure U.S. Production Capacity of Semiconductor-Grade Polysilicon," press release, January 7, 2025, https://www.commerce.gov/news/press-releases/2025/01/department-commerce-announces-chips-incentives-award-hemlock.

22. Yuan Gao, "China Creates $47.5 Billion Chip Fund to Back Nation's Firms," *Bloomberg*, May 27, 2024, https://www.bloomberg.com/news/articles/2024-05-27/china-creates-47-5-billion-chip-fund-to-fuel-self-resilience.

23. See Dylan Butts, "Japan Is Ramping Up Efforts to Revive Its Once Dominant Chip Industry," *CNBC*, November 13, 2024, https://www.cnbc.com/2024/11/13/japan-is-ramping-up-efforts-to-revive-its-once-dominant-chip-industry-.html. Also see Takashi Mochizuki and Yuki Furukawa, "Japan Earmarks Another $5.4 Billion for Chip Startup Rapidus," *Bloomberg*, March 30, 2025, https://www.bloomberg.com/news/articles/2025-03-31/japan-earmarks-another-5-4-billion-for-chip-startup-rapidus.

24. Jillian Deutsch, "EU Nations Advance €43 Billion Plan to Become Semiconductor Hub," *Bloomberg*, November 23, 2022, https://www.bloomberg.com/news/articles/2022-11-23/eu-nations-advance-43-billion-plan-to-become-semiconductor-hub.

25. Kim Tong-hyung, "South Korea to Boost Support of Semiconductor Industry in the Face of Trump's Tariffs," *Associated Press*, April 15, 2025, https://apnews.com/article/south-korea-semiconductors-financial-support-trump-tariffs-658f4ca0007ebef98d2b477ecd531867. See also Peterson Institute for International Economics,

"The US and Korean CHIPS Acts Are Spurring Investment but at a High Cost," June 10, 2024, https://www.piie.com/blogs/realtime-economics/2024/us-and-korean-chips-acts-are-spurring-investment-high-cost.

26. National Science and Technology Council, "Taiwan Chip-Based Industrial Innovation Program," November 13, 2023, https://english.ey.gov.tw/News3/9E5540D592A5FECD/e4e0680d-0ca2-4239-9e14-34716721366f.

27. Associated Press, "Britain Unveils $1.2B Strategy to Boost Computer Chip Industry," *AP News*, May 19, 2023, https://apnews.com/article/computer-chips-semiconductor-strategy-britain-75a38608b059773f0c1b130c482df84e.

28. McKinsey and Company, "How Semiconductor Companies Can Fill the Expanding Talent Gap," accessed July 22, 2025, https://www.mckinsey.com/industries/semiconductors/our-insights/how-semiconductor-companies-can-fill-the-expanding-talent-gap.

29. Speech by US Secretary of Commerce Gina Raimondo titled "The CHIPS Act and a Long-Term Vision for America's Technological Leadership" at Georgetown University's School of Foreign Service, https://www.commerce.gov/news/speeches/2023/02/remarks-us-secretary-commerce-gina-raimondo-chips-act-and-long-term-vision.

30. Semiconductor Industry Association and Oxford Economics, "Chipping In."

31. Boston Consulting Group and Semiconductor Industry Association, "Strengthening the Global Semiconductor Supply Chain in an Uncertain Era," April 2021, https://web-assets.bcg.com/9d/64/367c63094411b6e9e1407bec0dcc/bcgxsia-strengthening-the-global-semiconductor-value-chain-april-2021.pdf.

32. SEMI, "More than Half of American-Owned IC Fab Capacity Exists Overseas," press release, accessed July 22, 2025, https://www.semi.org/en/news-resources/press/knometa-american-ic-manufacturers-overseas-capacity.

33. Marie Garcia Bardon and Bertrand Parvais, "The Environmental Footprint of Logic CMOS Technologies," IMEC, December 17, 2020, https://www.imec-int.com/en/articles/environmental-footprint-logic-cmos-technologies.

34. Naga Chandrasekaran, "Intel Foundry Direct Connect," presentation, Intel Foundry, April 29, 2025.

35. Kenneth Flamm and William B. Bonvillian, "How Intel's Innovation Problem Became a National Security Crisis," *American Affairs* 9, no. 1 (Spring 2025), https://americanaffairsjournal.org/2025/02/how-intels-innovation-problem-became-a-national-security-crisis/; Kenneth Flamm and William B. Bonvillian, "Solving America's Chip Manufacturing Crisis," *American Affairs* 9, no. 2 (Summer 2025), https://americanaffairsjournal.org/2025/05/solving-americas-chip-manufacturing-crisis/.

36. Moore's Law is the empirical observation that the number of transistors on a microchip doubles approximately every two years while processing costs are halved, originally formulated by Intel cofounder Gordon Moore in 1965.

37. Samsung, "Samsung Joins ASML's Customer Co-Investment Program for Innovation, Completing the Program," press release, August 27, 2012, https://www.sec.gov/Archives/edgar/data/937966/000119312512370598/d402810dex991.htm.

38. G. Dan Hutcheson, "SEMI ISS, Anne Kelleher, Moore's Law," *Tech Insights*, accessed July 24, 2025, https://www.techinsights.com/blog/semi-iss-anne-kelleher-moores-law.

39. Jesus A. del Alamo, Dimitri A. Antoniadis, Robert G. Atkins, Marc A. Baldo, Vladimir Bulovic, Mark A. Gouker, Craig L. Keast, et al., "Reasserting US Leadership in Microelectronics: A White Paper on the Role of Universities," 2021, https://dspace.mit.edu/handle/1721.1/139740.

40. McKinsey and Company, "How Semiconductor Makers Can Turn a Talent Challenge into a Competitive Advantage," September 7, 2022, https://www.mckinsey.com/industries/semiconductors/our-insights/how-semiconductor-makers-can-turn-a-talent-challenge-into-a-competitive-advantage.

3
Biomanufacturing: Transforming US Capabilities and Capacity to Accelerate the Domestic Bioeconomy

J. Christopher Love

Strategic Importance

The global bioeconomy has the potential to generate more than $4 trillion in revenue by 2035, with nearly $1 trillion in the United States, based on a valuation in 2025 of approximately $240 billion.[1] Indeed, estimates suggest that biological sources could generate up to 60 percent of the materials used in all products worldwide. Biologically enabled products could emerge in virtually all sectors, including agriculture and food production, health care, defense, and consumer goods. Bio-enabled domestic production can increase US national security and resilience, for example, through greater biopharmaceutical drug, food, and water security.[2] For the United States to lead in the bioeconomy, it needs to transform its domestic base for biomanufacturing so it can support the commercialization of all forms of biologically enabled products. We are far from achieving that now. Not only are US innovative bioindustrial and biopharmaceutical companies seeking overseas capacity for scaling up production;[3] many countries, including China, India, South Korea, France, and Canada,[4] are committing to invest billions to create the required infrastructure for biomanufacturing.

Increasing domestic biomanufacturing capabilities and capacity not only would generate benefits for the economy as a whole, but also would strengthen US capability to respond rapidly to emerging health threats and other concerns for national security.[5] Fundamental discoveries and applied innovations in both how and where we perform biomanufacturing are needed in every sector poised to benefit from the bioeconomy: bioindustrial chemicals, new materials and foods, reagents for diagnostics, novel pharmaceuticals and vaccines, and essential intermediates and raw materials used for products that only biology can produce (such as artificial tissues and "organs-on-a-chip" for reducing animal testing).

Empowering the US bioeconomy can also help reduce domestic demands for energy while bolstering a sustainable domestic infrastructure for manufacturing. Biologically based manufacturing can use renewable feedstocks and alternative routes to biodegradable products like plastics, decreasing reliance on nonrenewable sources and complex supply chains. By enabling greater innovation and scale in the US biomanufacturing sector, the production of materials by biological processes could reduce CO_2 equivalent emissions by up to 90 percent if realized at scale.[6] Innovations in biomanufacturing methods for producing goods, particularly with the growing impact of artificial intelligence, could also reduce the amount of energy consumption, water usage, and waste in manufacturing, modernizing manufacturing for some chemicals and enabling more resilient and secure supply chains.

Current Landscape

The United States is a global leader in biotechnology innovations but lacks the manufacturing infrastructure and expertise necessary to bring these ideas to the market at the same pace as it generates innovative new products. Advances in synthetic biology and bioengineering pioneered in the United States face a dearth of domestic manufacturing capabilities and facilities that threatens to shift new technologies and related manufacturing overseas, jeopardizing US leadership and jobs in this area.[7]

The breadth of products enabled by biotechnology is broad—ranging from small-molecule and commodity goods like new materials for clothing and nutritional supplements for foods to high-value biopharmaceutical medicines for cancer and health. The United States has led in technology innovation across this entire range because of the strength of its academic research and the availability of private risk capital to invest in startup companies focused on the invention of biotechnology products. But these innovative US companies find it difficult to gain access to the domestic manufacturing capacity and capabilities needed to bring their products to market. The time and capital expense required to build manufacturing facilities (as currently designed) cause many US startups to seek production at existing facilities overseas.

In contrast, many governments, including those of China, India, South Korea, France, and Canada,[8] are committing to investing billions to create the required infrastructure for biomanufacturing. Indeed, the Chinese government's strategic decision to invest in operating contract manufacturing companies over the last twenty years has grown their capacity for biomanufacturing robustly for both medical and nonmedical applications, leading innovative US companies to depend on Chinese facilities for early development and commercialization.[9] For example, a recent survey of US biopharmaceutical companies highlighted that about 80 percent rely on Chinese biomanufacturers.[10]

The cost of building and operating manufacturing facilities depends on the volume of production (grams to tons) and the market price of the products (cents per ton to dollars per gram). For example, replacing high-volume, low-margin commodity chemicals with biologically derived ones involves different scaling challenges than those for low-volume, high-margin products like biopharmaceuticals. These two bookend cases are justifiably the focus of two

Manufacturing USA institutes, BioMADE and NIIMBL, respectively.[11] But these institutes do not address many critical aspects of biomanufacturing, including critical biomaterials (such as proteins, DNA, and RNA) used as reagents, additives, intermediates, and food materials. Emerging critical areas for the bioeconomy—such as the production of bio-based textiles, fungi-based materials for non-plastic packaging and construction, and protein-based ingredients for food applications, among other applications—represent a distinct challenge for manufacturing.[12] These classes of products can be manufactured only with biology, and there is a significant gap in basic knowledge of how to produce them affordably and sustainably.

All biomanufacturing inherently faces two basic challenges: (1) living biological systems are not (yet) as predictable as chemical synthesis, and (2) the complexity of a biological product links directly to its quality and function. That is, many of the desired features of the biological product depend on its inherent complexity, but that complexity also introduces many ways in which the quality or function of the product can vary depending on how it is produced, stored, and transported.

If the United States is to become a global leader in the bioeconomy, it will need to make many advances in the science of biomanufacturing and the implementation of new capabilities. At the same time, biomanufacturing draws from the broader domestic industrial ecosystem, and the United States will face the typical challenges and opportunities around developing the workforce, building a strong domestic supply chain, and financing manufacturing scale-up.

Gaps and Opportunities

Availability of Manufacturing Facilities

The availability of manufacturing facilities in the United States, especially state-of-the-art facilities, is woefully insufficient, given the exponential growth in biological innovation. New ideas for biologically derived products are being generated by academic centers, national labs, and private startups in the United States far more quickly than its manufacturing capacity is increasing.[13] New manufacturing facilities often require two to five years to build and up to $2 billion of capital investment. Companies seeking to introduce innovative new products cannot afford such costs or such delays. Contract manufacturing organizations (CMOs) cannot solve this problem of biomanufacturing capacity because they also face similar capital constraints and do not have sufficient incentives to invest in transformative innovation. Furthermore, the business models for CMOs often rely on a small number of large contracts to maintain their operations in their existing facilities rather than servicing many new innovative companies.

Simply building more facilities and capacity, however, cannot match the pace of innovation and formation of hundreds of new companies annually that need biomanufacturing capacity.[14] For example, one biomanufacturing facility may operate only 20 to 40 times per year at peak efficiency to produce fewer than 20 to 40 different products (or even just a single one). This mismatch introduces long delays in the commercial development of new products and increases the costs of delivering them to the market. If the United States continues on its current trajectory, there will always be capacity issues in the country that limit the ability of innovators to scale production of new products in the United States.

The opportunity for the United States is to create and scale new types of manufacturing facilities.[15] Innovative companies need affordable access to smaller, faster, and more cost-efficient biomanufacturing facilities. In order for such facilities to emerge in the United States, they need to be 10 to 100 times lower in cost (roughly $10 million to $100 million) and time (one to three years) than the cost and time of setting them up and operating them overseas. Operating today's biomanufacturing processes requires significant financial investment and highly specialized equipment, and these requirements pose a serious barrier for investors transitioning products to the market. Hence, as noted above, companies today rely primarily on overseas manufacturing capacity for initial market entry. Advances in automation technologies and biological engineering, however, have resulted in advanced prototypes and early commercial development of "small footprint" continuous manufacturing systems for proteins, viral therapies, cells, high-value small molecules (e.g., fragrances, drugs), and other bioproducts. Such systems show the potential for a different approach to expanding manufacturing capacity than the current one.

Combining "all-in-one" types of equipment with modular, "smart" manufacturing clean rooms makes it possible to build small manufacturing facilities virtually anywhere, including in warehouses and labs, at significantly lower capital and operating costs—at a fraction of the cost of current state-of-the-art facilities. Such facilities could allow companies to build their own manufacturing in the United States. These manufacturing innovations could also reduce energy, water demands, and waste by as much as 90 percent compared with traditional biomanufacturing.[16]

Continuous manufacturing technologies are another means of reducing the requirements for scale-up since increasing production volumes can be accomplished by extending the time of production, making it feasible to "scale out" production rather than scaling up and realize efficiencies in facilities for both pilot and commercial manufacturing.

Such "leapfrog" manufacturing technologies and approaches have been demonstrated but are still in their adolescence with respect to adoption and implementation. Advanced biomanufacturing capabilities were funded by the Department of Defense under the DARPA/Joint Program Executive Office for Chemical, Biological, Radiological and Nuclear Defense (JPEO-CBRND) from the early 2010s to the early 2020s. These new biomanufacturing technologies could benefit many sectors from defense to health security, but few have matured into routine commercial applications with Technology Readiness Levels (TRLs) of 8–9. (TRL is a common metric using a 1–9 scale to describe how close a technology is to regular commercial deployment.[17])

These technologies for new manufacturing systems need to be accessible in the United States so that companies can gain experience in learning how to use them, enabling them to grow to commercial scale. The United States—a global leader in biopharmaceutical development and manufacturing—could also better understand and apply lessons learned from continuous manufacturing in biopharmaceuticals to other related sectors (e.g., materials for consumer products, regulated new protein supplements, and nutritional additives).

Need for Bespoke Processes to Manufacture Biologically Derived Products

Most biologically derived products currently require bespoke processes for manufacture. A biomanufacturing facility optimized today for one type of product is usually not compatible with other new types of products without significant additional capital expenses and time. This constraint is due to the complexity of biologically derived products such as insulin, protein products, mRNA vaccines, cell therapies, and fine chemicals. This complication means that building more capacity does not equate directly to increased productivity; some capacity simply won't accommodate some products, and additional processing may be required on a product-by-product basis to assure safety and quality, and to meet cost targets.

Creating national capabilities for routine 100-day response times in pandemics or other health emergencies will depend on having well-established platforms and design rules (called Design for Manufacturing [DFM] in some industries) for producing multiple different types of interventions, as was evident in the COVID-19 response with both mRNA and monoclonal antibodies. A robust manufacturing base for a range of bio-enabled products from reagents and materials to health products is a core strategic need for domestic resilience in supply chains and reduced dependencies on imported raw materials. Unfortunately, there are very few examples globally of biomanufacturing processes and facilities capable of producing a wide range of products within a class or type of product.[18]

There are some examples of manufacturing processes that are more broadly applicable because of standardization of biomanufacturing processes. The manufacturing of monoclonal antibodies as biopharmaceuticals is the most developed example of a highly standardized and generally transferable platform used across the industry. Recent developments of manufacturing platforms for mRNA vaccines represent another. There is an opportunity to learn how and when to apply this model to bioindustrials and other critical biomaterials used in consumer goods and foods.

These manufacturing platforms were established over time through a combination of pre-competitive academic-industrial collaboration and government support.[19] The similarities among the target classes (antibodies or mRNA) mean it is feasible today to create facilities for producing many similar products within the class (for cost savings) and to transfer processes from one facility to another with minimal overhead (for resiliency).

Other critical classes of products for national supply chains, such as protein enzymes, plasmid DNA, small proteins like insulin, and certain emerging classes such as cell and gene therapies, are amenable to similar innovations in the biomanufacturing industry, but those innovations either are not fully developed or do not exist at all.

Lessons learned from the biopharmaceutical industry in standardization and alignment of biological processes for products like monoclonal antibodies and mRNA could help inform new strategies for related product classes. For example, high-quality protein production for other applications (such as nutritional supplements, advanced materials, sensors, and diagnostics) or cell manufacturing could benefit from the biomanufacturing learning curve in biopharmaceuticals.[20]

Accelerate Adoption by Industry of Disruptive Biomanufacturing Technologies

Companies are generally slow to invest in or adopt new biomanufacturing technologies due to concerns about readiness for use, impact on existing capital investments, or perceptions of regulatory risks. As a result, most disruptive technologies for biomanufacturing are advancing in academia, startups, or small- and medium-sized enterprises (SMEs). But timely access to financial resources (including commitments for purchase of products or services) is required to develop and deploy innovative manufacturing technologies.

The most disruptive technologies will emerge from combining new technologies synergistically rather than through the more common path today in the United States of incremental replacement of existing manufacturing steps.[21] As a result, significant investments and research to date on potentially transformative projects in biomanufacturing methods and approaches have not matured as rapidly as the underlying advances in technology (automation, bioengineering, software/AI) could enable.

The solution lies in de-risking emerging "leapfrog" technologies in early commercialization through RD&D investments. Since the early 2000s, the National Science Foundation (NSF) and Department of Defense (DoD) have invested modestly in fundamental science—about $100 to $500 million cumulatively—to advance new concepts for efficient, automated, and sustainable biomanufacturing, including continuous fermentation, novel purification approaches, data-driven predictions of bioprocesses, and automated process controls. Many of these technologies, however, have languished in academic centers or have transitioned into startups or SMEs, needing further validation to promote adoption by larger companies. Such technologies could help address critical challenges in scaling the manufacturing of critical raw materials for the bioeconomy, such as proteins, cells, and DNA, but there are limited opportunities for showcasing the potential of these technologies to attract further private-sector investment through risk capital or large corporations.

Funding commitments from multiple federal agencies, or advanced purchasing commitments for startups and SMEs, are necessary to help pull advanced biomanufacturing technologies through to commercialization. Given the rapid development of bio-enabled products and the lagging biomanufacturing capacity and capability in the United States, it is urgent that programs be put in place to de-risk and establish holistic new solutions. These could be established through grand challenges or similar transformative mechanisms issued by 2029. Given the commitments already in place by other countries seeking to establish their own leadership, the United States needs to accelerate its efforts in this area.

Recommendations

1. Invest in the science of biomanufacturing.

The United States remains a leader in scientific breakthroughs in biology, including synthetic biology; we need a similar innovation engine and accelerated pace for transforming biomanufacturing, where the United States is lagging.

- The United States does not currently have enough investment in biomanufacturing in the basic research stage (TRL 1–3) nor in the early-stage scale-up phases (TRL 4–8) to sufficiently pull private-sector investment into biomanufacturing.[22] In order to realize the opportunity to dominate this field, the federal government should establish programs to encourage investment in innovative biomanufacturing processes and facilities, including mechanisms to provide strategic capital (e.g., low-interest federal loans and/or working capital, procurement commitments). These programs should promote biomanufacturing research, development, and demonstration (RD&D), focusing on disruptive biomanufacturing-directed research (TRL 1–3) with a commitment to transition through later stages (TRL 4–8). The latter stages can demonstrate use cases (e.g., prototype support for pilot product introduction, first-in-human clinical trial, life cycle analysis for sustainability).

- More specifically, the government should launch two to three transformative large-scale initiatives designed to change the current trajectory of biomanufacturing in the United States by 2029 through mechanisms like those that have been used for investment at scale in other critical industries for the country. Potential sponsors include the DoD, the Advanced Research Projects Agency for Health (ARPA-H), the Food and Drug Administration (FDA) and other parts of Health and Human Services (HHS), the Department of Energy (DoE), the US Department of Agriculture (USDA), and the National Science Foundation (NSF). The initiatives should work toward the following goals:
    - Build sustainable manufacturing solutions for ultra-low-cost, high-quality proteins or cells ($1 per kilogram) at 100-kilogram scales to enable regional commercial development in the bioeconomy.[23]
    - Validate "platform" or "standardized" production processes for multiple products within classes of input materials (e.g., proteins, cells, viruses) critical for the US supply chain in the bioeconomy or enabling for a 100-day response with a new vaccine, for example.

    - Create data-sharing platforms and artificial intelligence/machine learning (AI/ML)-based design tools to accelerate biomanufacturing process design[24] and implementation in order to reduce time and costs for transitioning ideas to market-ready products. AI/ML tools could also aid in transitioning TRL 1–3 technologies into later development stages.
- Existing public/private collaborations in biomanufacturing, such as NIIMBL and BioMADE, do not currently have sufficient resources or mandates to meet the full scope of their potential in boosting the US biomanufacturing sector. The federal government should provide additional funding to charge Manufacturing USA Innovation Institutes such as NIIMBL and BioMADE with additional goals, focusing on TRL 4–7, such as (1) advancing holistic new or disruptive technology solutions for sustainable and affordable biomanufacturing, including integration of complementary technologies as needed,[25] and (2) reducing/offsetting the actual costs of participation for SMEs by changing intellectual property (IP) terms (e.g., shared revenue pools for cross-licensed IP) and cost-matching requirements (tax credits for costs incurred similar to internal R&D).
- Federal agencies should invest in creating shared tools, resources, and systems to allow for data sharing and modeling in biomanufacturing. AI/ML tools require substantial (and differentiated) data for training and predictions, but manufacturing data by its nature has minimal variations. To achieve sufficient scale as a nation, data sharing of process conditions and outcomes is critical. Policies to promote open data sharing in manufacturing and implementation of certain approaches to data modeling (e.g., federated learning models) could balance competitive advantage with essential resources for national security and resilience in biomanufacturing. Interagency discussions on regulatory efficiency and timeliness should be

encouraged to meet the growing pace of innovation in biologically derived products.

- The federal government, philanthropic foundations, and nonprofit institutes should commission studies specific to biomanufacturing on (1) alternative business models and technology risk assessments for enabling affordable and available biomanufacturing capacity, and (2) cross-sector analyses to understand opportunities for technology adoption and transfer based on use cases (e.g., continuous manufacturing practices in pharmaceuticals for applications in precision fermentation). Such studies would help catalyze new biotechnology-enabled businesses and transfer learnings on best practices across sectors of the bioeconomy where common innovations in biomanufacturing practice and technologies are needed.[26]

2. Invigorate a national innovation pipeline for advanced biomanufacturing.

The United States needs to build capacity through regional bioeconomy innovation hubs around the country through strategies such as the following:

- Future government-sponsored capacity should be focused on supporting regionally distributed pilot hubs and early commercialization facilities. These facilities should not operate only as service providers, such as traditional contract manufacturing organizations (CMOs), as government-backed CMOs in the United States and in other parts of the world have not met the evolving needs of the biomanufacturing sector.[27] Instead, these facilities should strive to transfer biomanufacturing capabilities and expertise to regional industries.[28] Activities could include serving as innovation hubs for startups and economic development, emphasizing flexibility and accessibility of new technologies, and the exploration of novel biomanufacturing business models.[29]

 - Support for regional hubs of sustainable and interdependent bioindustrial and bio-specialty production could include education, workforce development, training facilities, and capacity. These hubs should have the space, advanced technology, and incentives to prioritize innovation in biomanufacturing. They could be modeled after the Commerce Department's Regional Technology and Innovation Hubs created under the CHIPS and Science Act.[30]
 - The development of the regional craft brewing industry is an interesting model for developing regional specialization and growth of performing SMEs with independent manufacturing capabilities and robust workforce for all levels.[31]
- Federal agencies should identify bio-related goods and products that rely on advanced biomanufacturing technologies and require some domestic production capacity for national security purposes (e.g., medicines; critical biomaterials such as DNA, RNA, and proteins; and nutritional components). Agencies should develop a strategy for domestic production, potentially using advance market commitments with incentives for incorporating disruptive manufacturing solutions.
- The federal government should work with states, local governments, nonprofit organizations, technical and community colleges, and universities, along with philanthropic foundations and other NGOs, to promote education and training from K–12 to post-graduate education to raise awareness of the vital importance of biomanufacturing for addressing current and future national economic growth and sustainability.
 - Education could include funding additional university centers of excellence with outreach programs (e.g., NSF Engineering Research Centers or similar mechanisms) and creating innovation awards (e.g., NIH Pioneer, NSF TRAILBLAZER, DoD Vannevar Bush Fellowships) and funds to

support new and outstanding investigators around the country working on the science of biomanufacturing.

3. Create financial incentives for biomanufacturing investment. Financial incentives are needed to motivate startups and SMEs to invest in biomanufacturing.

- Investment tax credits would help incentivize startups and SMEs to invest in developing their own biomanufacturing facilities and expertise, which in turn would allow for greater iteration and innovation in the biomanufacturing process. Tax credits could also provide more incentive to industry to partner on adopting and building advanced manufacturing capabilities and capacity. The early success of investment tax credits in encouraging private-sector investments in the semiconductor and clean energy industries in the United States could provide a model for biomanufacturing.
- Federal programs should provide low-interest, long-term loans and other forms of capital to SMEs to support manufacturing equipment development, software/automation, and building out capacity. Several existing programs could be explored for this, including HHS BARDA Ventures, In-Q-Tel, and the EXIM Bank's Make More in America program. The DoD's new Office of Strategic Capital is also an important resource to ensure US superiority in key areas of biomanufacturing.

Conclusion

Today, the United States is not positioned to lead in the bioeconomy.[32] We are underinvested in capital, infrastructure, and people. There is an urgent need to realize disruptive "leapfrog" solutions to align our manufacturing resilience with our pace of innovation and to create a sustainable future. The current national strategy for assuring timely, sustainable capabilities in innovative biomanufacturing is unclear and assumes that ongoing needs in biomanufacturing can be met with current technology. The convergence of biological engineering with the fast-emerging capabilities in artificial intelligence offers a unique opportunity for the United States to accelerate its bioeconomy. A well-defined national strategy, however, will be required in order to establish the requisite technologies, business structures, and financial tools to promote the change in the biomanufacturing base that is needed to assure the country's leadership in this next industrial revolution.

The author would like to thank the following colleagues for their contribution to this chapter: Catherine Cabrera, Group Leader, Biological & Chemical Technologies Group, MIT Lincoln Laboratory; Hannah Frye, Senior Policy Advisor, MIT Washington, DC, office; Paula Hammond, Vice Provost and Institute Professor, MIT; and Kristala Prather, Arthur Dehon Little Professor, Professor of Chemical Engineering, Head, Department of Chemical Engineering, MIT.

Notes

1. The bioeconomy emphasizes products and services that use biological resources like plants, animals, and microorganisms as renewable resources. A. Hodgson, J. Alper, and M. E. Maxon, *The U.S. Bioeconomy: Charting a Course for a Resilient and Competitive Future* (Schmidt Futures, 2022), https://doi.org/10.55879/d2hrs7zwc. See also Biotechnology Innovation Organization (BIO) and Kearney, "Projected Impact and Growth of a Fully Unleashed Bioeconomy," March 2025, https://www.bio.org/sites/default/files/2025-03/Bioeconomy%20Impact%20Modeling%20Final%20Report_March%202025_1.pdf.

2. White House Office of Science and Technology Policy, "Bold Goals for U.S. Biotechnology and Biomanufacturing: Harnessing Research and Development to Further Societal Goals," March 2023, https://bidenwhitehouse.archives.gov/wp-content/uploads/2023/03/Bold-Goals-for-U.S.-Biotechnology-and-Biomanufacturing-Harnessing-Research-and-Development-To-Further-Societal-Goals-FINAL.pdf.

3. Matt Blois, "The US Aims to Close Its Fermentation Capacity Gap," *C&EN Global Enterprise* 101, no. 9 (2023): 15–17, https://doi.org/10.1021/cen-10109-feature1; Biotechnology Innovation Organization, "BIO Survey Reveals Dependence on Chinese Biomanufacturing," May 9, 2024, https://www.bio.org/gooddaybio-archive/bio-survey-reveals-dependence-chinese-biomanufacturing.

4. PCAST, *Biomanufacturing to Advance the Bioeconomy*, executive summary.

5. National Security Commission on Emerging Biotechnology, "Charting the Future of Biotechnology: An Action Plan for American Security and Prosperity (Report Summary)," April 2025, https://www.biotech.senate.gov/wp-content/uploads/2025/04/NSCEB-Full-Report---Digital--4.28.pdf.

6. Boston Consulting Group and Synonym, *Breaking the Cost Barrier in Biomanufacturing*, February 2024, https://web-assets.bcg.com/b6/15/6a10d22c481e8bebaf0c2fab8294/bcg-breaking-the-cost-barrier-on-biomanufacturing-rev.pdf.

7. Executive Office of the President, President's Council of Advisors on Science and Technology, *Biomanufacturing to Advance the Bioeconomy* (December 2022), https://bidenwhitehouse.archives.gov/wp-content/uploads/2022/12/PCAST_Biomanufacturing-Report_Dec2022.pdf.

8. PCAST, *Biomanufacturing to Advance the Bioeconomy*, executive summary.

9. Fitch Ratings, "US Biomanufacturing Plan Poses Minor Risks to Chinese Pharma CDMOs," September 27, 2022, https://www.fitchratings.com/research/corporate-finance/us-biomanufacturing-plan-poses-minor-risks-to-chinese-pharma-cdmos-27-09-2022.

10. Biotechnology Innovation Organization, "BIO Survey Reveals Dependence."

11. Manufacturing USA is a federally sponsored network of eighteen manufacturing innovation institutes established in 2014, including BioMADE (Bioindustrial Manufacturing and Design Ecosystem) and NIIMBL (National Institute for Innovation in Manufacturing Biopharmaceuticals), which focus on bioindustrial and biopharmaceutical manufacturing, respectively. See https://www.manufacturingusa.com.

12. New applications for biotechnology include advances for foods, nutrition, and materials that require scales of manufacturing and quality metrics that fall between industrial products and medicines. These traits impose new requirements for biomanufacturing efficiency and operations distinct from low-value biochemicals and high-quality biologic medicines.

13. Blois, "US Aims to Close Its Fermentation Capacity Gap."

14. PitchBook data indicates 600 venture capital deals in the US biotechnology sector in 2021.

15. J. Christopher Love, "Transforming Manufacturing to Continue Leading the Innovation of Biopharmaceuticals," *The Bridge* 55, no. 2 (Summer 2025): 8–17, https://www.nae.edu/File.aspx?id=338964.

16. Michael Eder, "Enhancing Sustainability in the Pharmaceutical Industry with Single-Use Technologies," Single Use Support, June 10, 2021, https://www.susupport.com/knowledge/manufacturing-processes/sustainability/enhancing-sustainability-pharmaceutical-industry-with-single-use-technologies; Melanie Ottinger, Irina Wenk, Joana Carvalho Pereira, Gernot John, and Stefan Junne, "Single-Use Technology in the Biopharmaceutical Industry and Sustainability: A Contradiction?," *Chemie Ingenieur Technik* 94, no. 12 (2022): 1883–1891, https://doi.org/10.1002/cite.202200105.

17. There are no current initiatives or projects at either NIIMBL or BioMADE focused on de-risking the readiness and demonstration of these emerging innovative approaches. Technology maturity is often measured using Technology Readiness Levels (TRLs), a standardized 1–9 scale developed by NASA. Early levels (1–3) represent basic research and proof-of-concept work, while the highest levels (8–9) indicate technologies that have been fully tested and proven to work reliably in real-world operational conditions, and are ready for widespread commercial use. See https://www.nasa.gov/directorates/somd/space-communications-navigation-program/technology-readiness-levels/.

18. The industry refers to such capabilities as "platforms" for manufacturing where a class of product type can be accommodated in the same facility/process. Most products, however, require bespoke processes and facilities today. Such "platforms" aid the speed to production and optimize the use of a given facility (reducing operating costs).

19. The Department of Defense has historically led programs incentivizing public/private partnerships in biomanufacturing, as this sector has critical implications for national security. See "DOD Selects Just – Evotec Biologics for Biomanufacturing Optimization," *Genetic Engineering and Biotechnology News*, June 25, 2024, https://www.genengnews.com/topics/bioprocessing/dod-selects-just-evotec-biologics-for-manufacturing-optimization/.

20. The need for improved standardization in biomanufacturing has also been flagged by the National Security Commission on Emerging Biotechnology (NSCEB) in their December 2023 Interim Report, which can be accessed at https://www.biotech.senate.gov/press-releases/interim-report/.

21. Integrated new technologies can "leapfrog" existing ones, saving time and resources, to deploy the necessary infrastructure. Cellphones are one example where many regions could avoid the costs of installing landlines. New biomanufacturing technologies could provide similar savings by displacing current approaches.

22. The Manufacturing USA institutes are not aligned by charter (cost-sharing requirements, IP terms, membership composition) or sufficiently financially resourced to facilitate pilot production of nascent transitioning technologies emerging from academia or startups. See chapter 6, on advanced manufacturing.

23. Achieving this transformation in cost and scale for these classes of products would activate new parts of the bioeconomy for nutritional supplements, personal care products, new food products, and very low cost biopharmaceutical manufacturing, promoting new competition and broadening access. Diverse domestic capacity can promote lower drug pricing. See, for example, Anurag S. Rathore and Faheem Shereef, "The Influence of Domestic Manufacturing Capabilities on Biologic Pricing in Emerging Economies," *Nature Biotechnology* 37, no. 5 (2019): 498–501, https://doi.org/10.1038/s41587-019-0116-0.

24. Executive Office of the President, President's Council of Advisors on Science and Technology, *Supercharging Research: Harnessing Artificial Intelligence to Meet Global Challenges* (April 2024), https://bidenwhitehouse.archives.gov/wp-content/uploads/2024/04/AI-Report_Upload_29APRIL2024_SEND-2.pdf.

25. As one example, expanding biomanufacturing capabilities to include alternative biological sources (e.g., anaerobic bacteria, microbial consortia, fungi, plant cells) beyond those in use in traditional biological fermentation (e.g., mammalian cells, certain yeast, aerobic bacteria) could synergize with simpler, more cost-efficient production processes for extremely low cost and high-volume manufacturing (e.g., Catherine B. Matthews, Chapman Wright, Angel Kuo, Noelle Colant, Matthew Westoby, and J. Christopher Love, "Reexamining Opportunities for Therapeutic Protein Production in Eukaryotic Microorganisms," *Biotechnology and Bioengineering* 114, no. 11 [2017]: 2432–2444).

26. The landscape of biotechnologies is vast (from chemicals to cells to biopharmaceuticals) and manufacturing and regulatory requirements vary across sectors. Nonetheless, the same underlying elements and technologies are used throughout. Innovations in one area (e.g., continuous manufacturing with automated process controls) can inform other strategies in other sectors, but these industries today are largely siloed. For example, bioindustrial manufacturing relies primarily on stainless-steel, batch processing for many applications, but continuous production with single-use technologies, like those used in biopharmaceutical manufacturing, can reduce the costs and resources required to build and operate a facility.

27. CMOs have also come under intensified scrutiny over their ability to successfully carry out federal biomanufacturing contracts. For example, in a high-profile case during the COVID-19 pandemic, Emergent BioSolutions had to destroy 400 million doses of Johnson & Johnson COVID vaccines due to contamination, and hid evidence of quality control problems from FDA inspectors. See Sheryl Gay Stolberg, Chris Hamby, and Sharon LaFraniere, "Emergent Hid Evidence of Covid

Vaccine Problems at Plant, Report Says," *New York Times*, May 10, 2022, https://www.nytimes.com/2022/05/10/us/politics/emergent-fda-vaccine-covid-contaminated.html.

28. PCAST, *Biomanufacturing to Advance the Bioeconomy*.

29. DoD-sponsored BioMADE announced in June 2023 that it is in dialogue with six states to build bioindustrial manufacturing capacity along with its confirmed project in Minnesota. See https://www.biomade.org/infrastructure.

30. US Economic Development Administration, "Regional Technology and Innovation Hubs (Tech Hubs)," https://www.eda.gov/funding/programs/regional-technology-and-innovation-hubs.

31. The craft brewing industry had a $77.1 billion economic impact on the US economy in 2023 and created 460,000 jobs, spurring revitalization of rural economies, particularly in small towns and communities (see Brewers Association, "Economic Impact," https://www.brewersassociation.org/statistics-and-data/economic-impact-data/).

32. National Security Commission on Emerging Biotechnology, "Charting the Future of Biotechnology: An Action Plan for American Security and Prosperity (Report Summary)," April 2025, https://www.biotech.senate.gov/wp-content/uploads/2025/04/NSCEB-Full-Report---Digital---4.28.pdf.

4
Quantum Computing: Enabling Long-Term US Leadership Through a Sustained and Bold Science and Engineering Agenda

William D. Oliver and Jonathan Ruane

Strategic Importance

Quantum computing is a transformative technology that has profound implications for competitiveness, national security, and society as a whole.[1] It will not replace the existing computing infrastructure (referred to as classical computing), but will be used for specific applications, most likely in a hybrid platform that draws on both classical and quantum computing.

More specifically, quantum computing has the potential to transform critical sectors by harnessing quantum mechanical properties to solve problems that are intractable for classical computers.[2] Key applications include accelerating drug discovery through precise molecular simulations, advancing materials science for energy technologies, and optimizing complex financial and logistics systems. The country that achieves quantum leadership will gain decisive advantages in these strategically important industries.[3]

Advances in quantum computing are driving the need to create digital encryption protocols to replace those that currently underpin today's cybersecurity, privacy systems, defense communication, and business transactions. While it will likely be many years before we have sufficiently capable quantum computers to execute effective attacks on current encryption systems, the concern is that bad actors will store today's encrypted information and decrypt it later.[4]

Working quantum computers exist today, but their scale is such that they have limited capabilities when compared with the enormous advances we have seen in classical infrastructure over the past fifty years. The next major milestone for the quantum community is to develop a quantum computer with sufficient scale to perform a commercially useful task that is impossible (in practical terms) for any classical computing system—even the most powerful supercomputers available today. This is known as quantum advantage. The achievement of quantum advantage will herald a new era in computation capabilities. It is estimated that the market size of quantum computing could be a $45–$131 billion globally by 2040, although as this broad range illustrates, it is notoriously difficult to accurately forecast the timing and economic impact of technologies at an early stage of development.[5]

The United States is at the fore in striving to achieve quantum advantage, but it is by no means certain that the United States will be the first to get there. Many other nations are also investing heavily in this space. In particular, quantum computing is an important battleground within a larger United States–China competitive landscape.[6]

Quantum computing applications such as breaking cryptographic security systems or developing new chemistries require much more powerful and sophisticated computing systems than what is needed for initial quantum advantage. The primary bottlenecks that exist today can only be resolved through advances in scientific discovery and fundamental engineering. Therefore, the major focus for the United States in quantum computing should be funding and facilitating science and engineering research as well as early-stage "ecosystem" building. This will all require public and private investment as well as collaboration across academia, government, industry, and the nonprofit sectors. The aim should be to build capacity and expertise in quantum computing in multiple regions across the country.

Current Landscape

Public-Sector Investment

Most advanced economy nations have published national quantum strategies outlining their ambitions and commitments to increase investments in the field. In September 2018, the United States set out a national approach to achieving the strategic goal of maintaining and expanding American leadership in quantum information science.[7] This was shortly followed by enabling legislation, namely, the National Quantum Initiative Act (NQIA),[8] which provided a framework for coordinating federal quantum activities, authorized strategic roles for key agencies, and called for increased collaboration between government, federal laboratories, industry, and universities.

Multiple agencies—civilian and military—are engaged in this effort, including through federally funded research development centers (FFRDCs).[9] Legislation was introduced in 2025 to amend the NQIA with the objective of prioritizing public/private partnerships for near-term quantum application development with a focus on pilots, demonstrations, and proofs of concept.[10]

The federal government has roughly doubled civilian spending on quantum science and engineering research since 2019 (figure 4.1). Quantum computing was the largest benefactor of this, although it declined marginally in 2023 and 2024. A key pillar of this investment was funding authorized by the NQIA, which created research centers and educational programs. The original 2018 act authorized $1.2 billion over five years (2019–2023). The reauthorization of this act is likely to be considered in the 119th

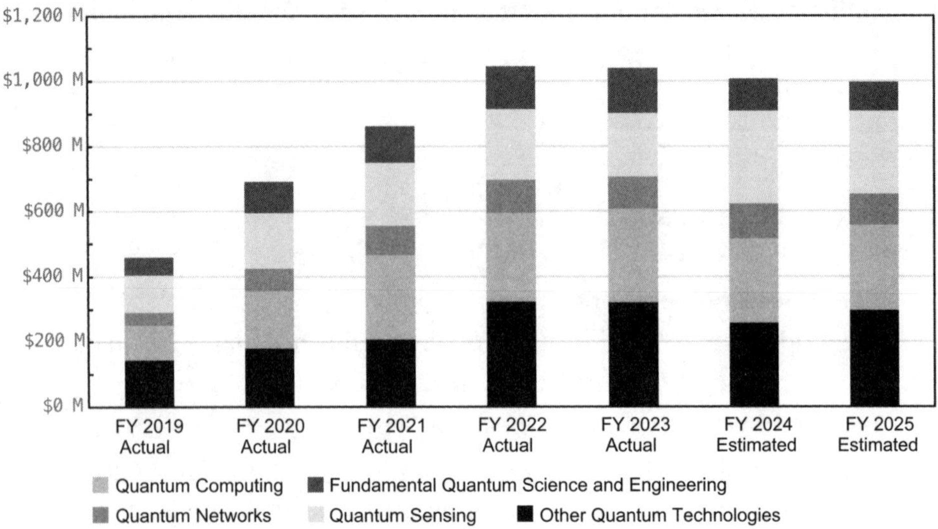

Figure 4.1

US Quantum R&D Budget Allocations by Component Area (2019–2025). Source: National Science & Technology Council, Subcommittee on Quantum Information Science, Committee on Science, National Quantum Initiative Supplement to the President's FY 2025 Budget (December 2024), https://www.quantum.gov/wp-content/uploads/2024/12/NQI-Annual-Report-FY2025.pdf, 8.

Congress.[11] Proposals aim to authorize appropriations totaling $2.7 billion over five years (2025–2029)[12]—a 125 percent increase. These funds would be directed toward key research agencies, including the Department of Energy (DoE), National Science Foundation (NSF), National Institute of Standards and Technology (NIST), and NASA.

Some analyses suggest that US public funding is dwarfed by China's, but reliable data is difficult to obtain. McKinsey reported over $15 billion in cumulative Chinese funding announcements across all quantum technologies over many years (figure 4.2). Whether this money has actually been deployed is unclear. Other countries, such as Germany, the United Kingdom, Japan, and South Korea, have also committed to scaling up their investments in quantum computing and have taken significant steps in that direction recently. In some cases, national and state governments are providing funds to US

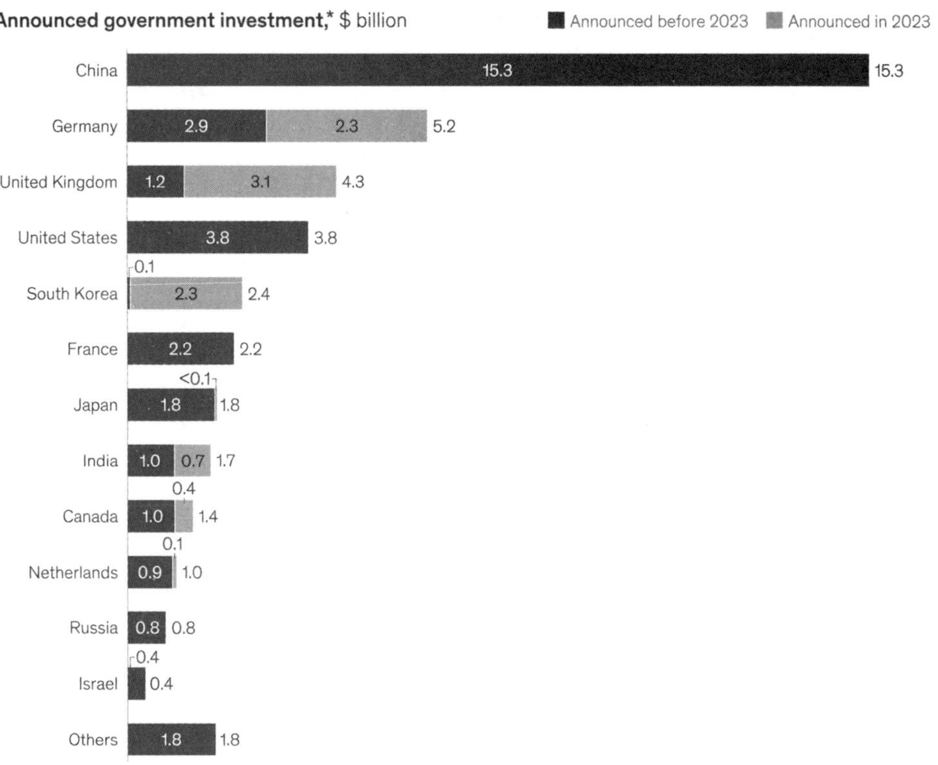

Figure 4.2
Global public announcements in public funding of quantum technology. From "Steady Progress in Approaching the Quantum Advantage," April 2024, McKinsey & Company, www.mckinsey.com. © 2025 McKinsey & Company. All rights reserved. Reprinted with permission.

startups to establish a footprint, such as incentives offered by Illinois and Australia to the US quantum computing firm PsiQuantum.[13]

Private-Sector Investment

While US public investment appears to lag China's, the United States appears to lead in private-sector investment. Gathering firm-level data internationally remains a challenge. Not all entities fully disclose their funding details, and challenges remain in terms of classification of funding types (e.g., equity vs. grants or loans). Investment in quantum divisions within large, publicly traded companies is not known, but these are some of the largest actors in the space. Our internal analysis suggests that US firms such as Google, IBM, Microsoft, and Amazon are by far the largest quantum investors of this type.

The best available data suggests that global venture capital investment in quantum startups exploded around 2020. Although it has plateaued in subsequent years, 2025 is likely to see further strong momentum. US firms represented almost half of global investment in quantum computing in 2024 (figure 4.3). This reflects the US's position as the leading nation in terms of private sector quantum investment.[14]

Despite the increase in investment in US quantum computing firms, the level of US private-sector investment in quantum computing is small compared with the investment in other sectors such as traditional software or AI, which received in excess of $100 billion in 2024.

Patents

The total number of quantum technology patent filings internationally grew five-fold across 2014–2024. China is the primary driver of this trend, and it is the dominant location for quantum patent filings, which grew by more than 600 percent during the same period. The United States has maintained second position during the same period. Although Chinese patents were historically seen as lower quality, many experts now suggest that the quality of the Chinese inventions is improving.[15] The quality of their quantum computing platforms is improving and, in many cases, competitive with the United States (figure 4.4).

Commercial Availability of Quantum Processors

Quantum processing units (QPUs) are the central hardware component of a quantum computer, analogous to the CPU in a classical

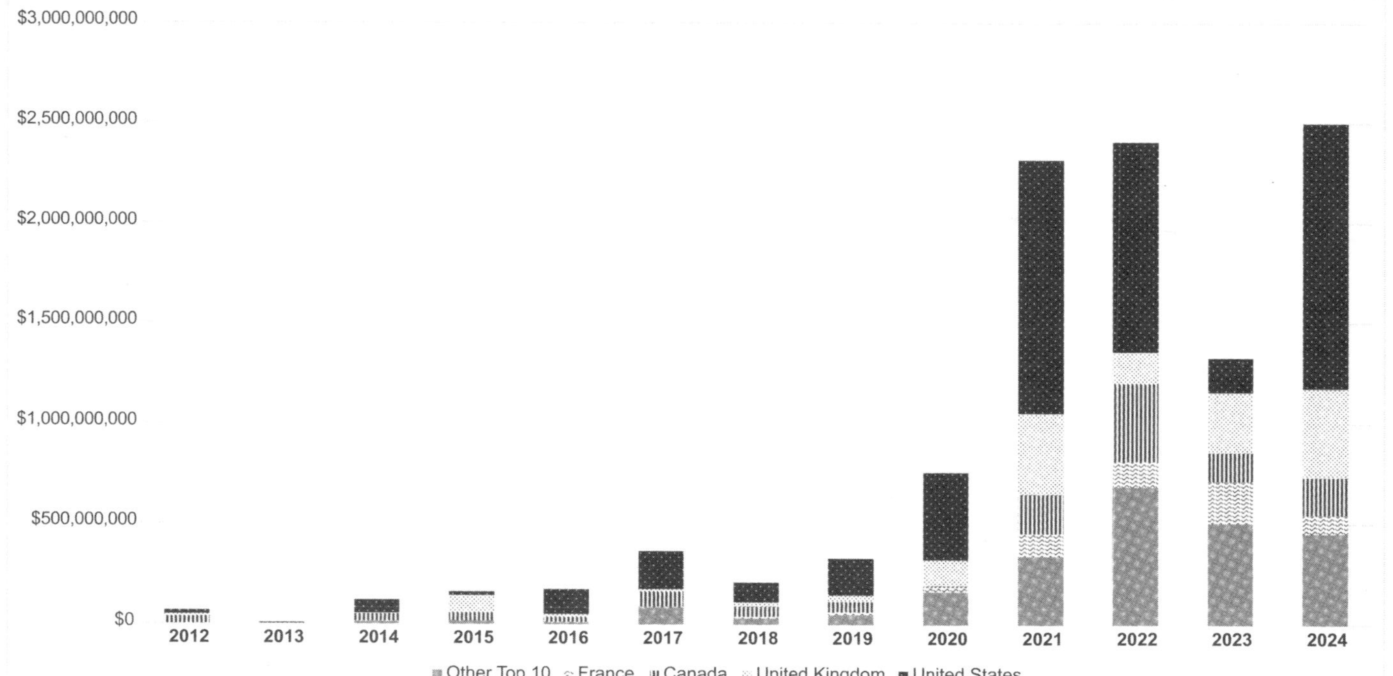

Figure 4.3

Quantum technology funding landscape by top ten countries (2012–2024). Adapted from Jonathan Ruane, Elif Kiesow, Johannes Galatsanos, Carl Dukatz, Edward Blomquist, and Prashant Shukla, “The Quantum Index Report 2025,” MIT Initiative on the Digital Economy, Massachusetts Institute of Technology, May 2025, https://qir.mit.edu/wp-content/uploads/2025/06/MIT-QIR-2025.pdf, 45.

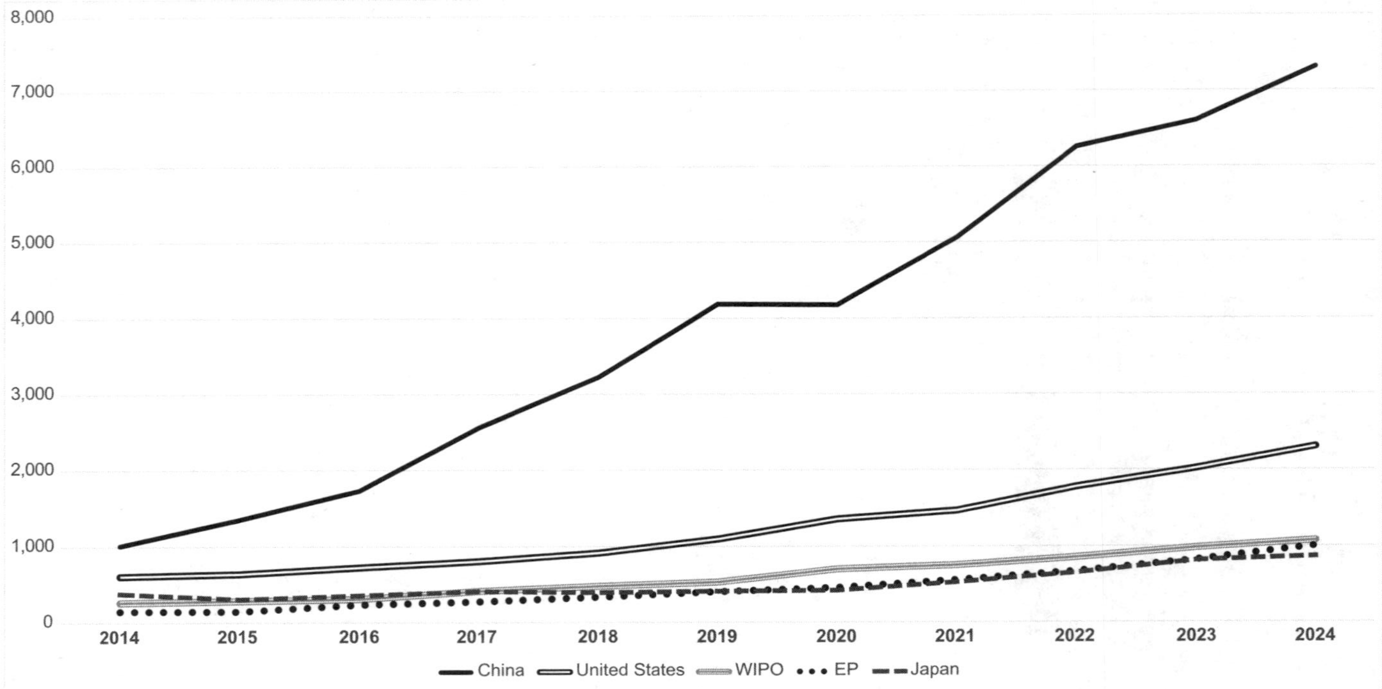

Figure 4.4
Quantum technology patents by country (2014–2024). Adapted from Ruane et al., "Quantum Index Report 2025," 21.

computer. As of 2025, approximately 40 QPUs are commercially available (public access via on-premises or cloud) from two dozen firms around the world. The type and performance of these QPUs vary, but the United States is the clear leader—China has 75 percent fewer QPUs available, and these devices are of lower performance.[16] These are useful metrics when comparing the capabilities of private-sector firms between nations, but more importantly, QPU availability is a vital facilitator in enabling ongoing use-case experimentation with industry (figure 4.5).

Gaps and Opportunities

Technology Scalability

For many quantum technologies today, the primary inhibitor to scaling is the insufficient maturity of the technology. This is exacerbated by the low cost and high performance of existing classical computers. For example, no quantum computer today is capable of outperforming a classical computer on any challenge that is commercially or practically useful. Breakthroughs in both fundamental science and fundamental engineering are needed to change this. Necessary technological advances that span physics, computer science, and engineering include creating quantum-specific algorithms, engineering improvements that allow for scaled-up system architectures, and developing new materials.[17] Achieving these advances will require investments across private companies, national research labs, and universities. While companies such as IBM and Google are engaged in the challenges entailed in actually scaling up current technologies, universities are doing the work in applied science and foundational engineering that will create the tools that will enable companies to scale in the future. Universities, of course, also educate the next generation of workers for research and industry.

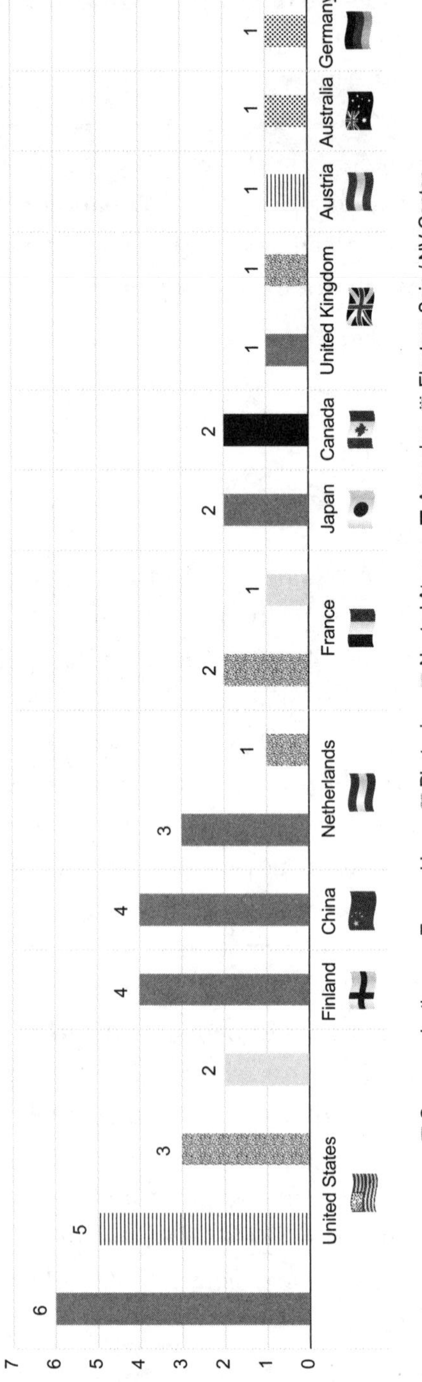

Figure 4.5
Commercially available QPU models per country. Source: Ruane et al., "Quantum Index Report 2025," 103.

US Funding Levels and Structure

As described above, both the federal government and the private sector have made significant investments in quantum computing in recent years. But the existing investments are not enough to rapidly propel the field. Quantum computing, like other "tough tech," requires sustained, patient capital. The United States must avoid a "quantum winter" in which funding for quantum research suffers a substantial downturn, where progress plateaus, and there is a cooling off in interest among scientists in the field. This would be somewhat akin to the "AI winters" in previous decades in which funding and interest in AI declined when hype gave way to disappointment in the progress in AI. The researchers who are vital to continuous development take at least a decade to train, and thus need sustained support through their training and beyond.

In addition, greater direction and focus are needed to get the most out of current US investments. Distributing money for individual labs to do "good science" is not enough. Funds need to be dedicated to achieving bold, overall applied science and foundational engineering goals, and then the various universities, in coordination with the US government and companies, need the funding stability and relative freedom to achieve those goals. Reducing regulatory burden and compliance requirements is key to saving money and personnel time so that they can be more efficiently applied to achieving research goals.

Developing a Talent Pipeline

The United States should aim to increase the development of its quantum researcher talent pipeline. The short-term demand for quantum experts may not be currently sufficient to generate the supply of scientists and engineers the United States will need in the future. While overall quantum job opportunities have increased since 2010 (figure 4.6), the field remains a relatively niche area in which there aren't large numbers of jobs. Anecdotally, the authors

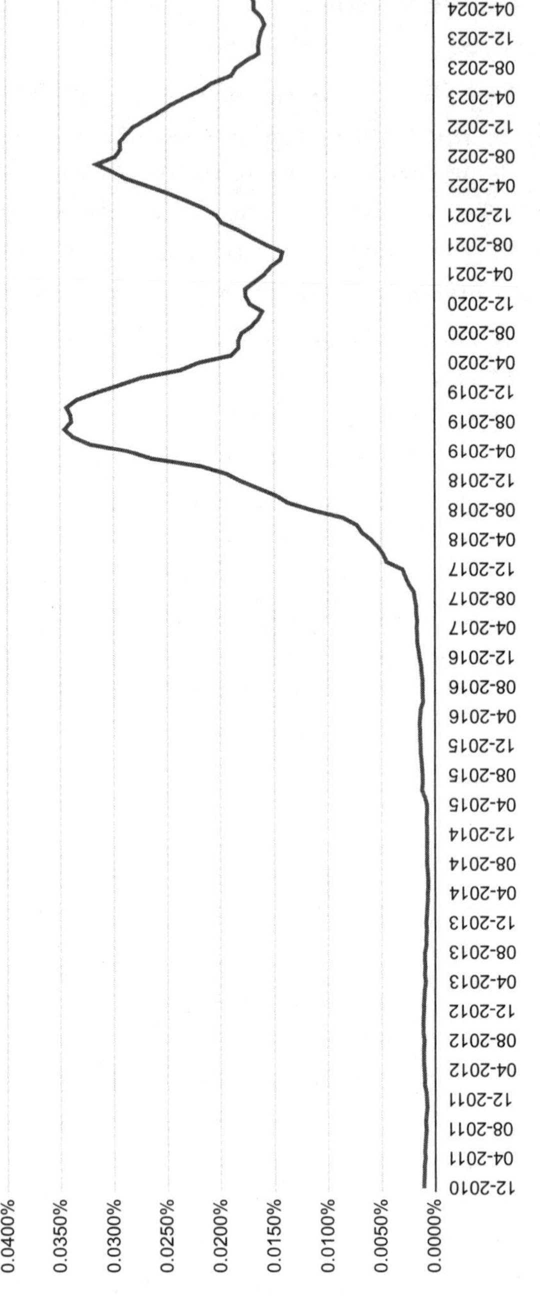

Figure 4.6
Percentage share of “quantum” mentions in US job postings (2010–2024). Source: Ruane et al., “Quantum Index Report 2025,” 68.

know that graduate students with quantum training are often lured away from the field by financial services firms that can pay much higher salaries. In addition, over time, the United States will need to have more focus on STEM at the K–12 level to eventually create the required flow of talent into graduate schools. More quantum education for undergraduates is also needed, as is worker (re)training. MIT, as well as other schools, has developed online professional development that targets people already in industry who want to pivot to quantum computing.[18]

Highly talented quantum researchers are rare even at a global level, and the United States must cast a wide net in order to build world-class capabilities. A flow of international students and workers is critical to provide the United States with the required quantum talent. The federal government should be making it easier for international students who receive their PhDs in the United States to remain here. The United States also has to do more to attract such students to begin with. In an American Physical Society survey of its international members, nearly half of the respondents who chose not to come to the United States said it was because they perceive the United States as unwelcoming.[19] More than 70 percent of foreign national respondents in the United States reported having had challenges obtaining a visa, with one in four reporting a delay of two months to one year.

Supply Chain Vulnerabilities

The key inputs to the quantum supply chain are evolving as the technology develops. A number of the inputs[20] have few suppliers, are expensive, and/or are concentrated in a few places. Europe and Asia are important supplier regions, including China, where lasers and rare-earth elements are currently sourced. Inputs like helium-3 gas are highly specialized, and the United States has experienced shortages in the past, requiring US government intervention.[21] Additional areas of supply concern include classical control electronics/field programmable gate arrays, classical control optics/lasers and cryogenic technologies, microwave control electronics, and optics.

To facilitate the development of a secure and resilient quantum supply chain, the US government will need to invest in analyzing, tracking, and understanding the market in more depth. The United States needs to know where vulnerabilities lie in the quantum supply chain and how to anticipate potential disruptions. The United States must prioritize building domestic capacity, especially for inputs that are critical to national security. The United States should work with allies and partners to try to prevent sole sourcing, particularly from China. In that regard, it is worth noting that quantum computing requires much of the same equipment as semiconductor manufacturing, so it may benefit from the recent investments in semiconductor production in the United States through the CHIPS and Science Act of 2022. For example, the Defense Department's Microelectronics Commons hubs could be enhanced to provide shared facilities for development and prototyping quantum devices and systems. The United States may want to consider how to create an IMEC-like center for quantum computing. IMEC was founded in 1984 as a nonprofit organization by researchers at KU Leuven and the Flemish government and is the world's largest independent research and innovation center for nanoelectronics and digital technology.[22]

Recommendations

1. Fund foundational research consistently over time.

Achieving the full promise of quantum computing is still a number of years away and, as noted above, requires significant breakthroughs in fundamental science and foundational engineering. The United States needs to leverage its world-class research universities to ensure it leads the way. The absolute size of the investment is important, but so is consistency of funding. The technology is still at an early stage of development and requires a stable, long-term commitment from policymakers and industry to ensure consistent progress is achieved. The federal government should commit to steady, predictable funding for quantum research. It takes at least a decade to train a PhD researcher in the field (undergraduate plus graduate education). If the United States fails to ensure continuous support for the sector in the coming years, it will be very difficult to regain lost momentum.

Specifically, the federal government should do the following:

- *Focus on coordinated efforts to achieve broader goals.* As noted earlier, coordinated efforts in fundamental science and fundamental engineering with clear goals are needed. Scientific research should focus on developing new quantum algorithms and applying existing algorithm primitives to real commercial problems to increase the utility of quantum computing.
- *Establish larger-scale goals.* Again, engineering is needed, not just science. Among other things, engineers need to develop scaling concepts and then test those developments at a larger scale of qubits (not necessarily at the largest scales). One specific focus of additional funding should be larger-scale quantum computing systems. As of 2025, only two government programs—IARPA ELQ (the Entangled Logical Qubits program at the Intelligence Advanced Research Projects Activity) and DARPA QBI (the Quantum Benchmarking Initiative at the Defense Advanced Research Projects Agency)—focus on larger-scale systems and error correction. (Note: This chapter does not deal with classified programs.) These kinds of larger-scale goals are needed to direct research and promote progress. Government investment is also needed in complementary capabilities that support research and development. One example is the expansion of

facilities that can fabricate experimental quantum chips. This will aid in the rapid turnaround of experimental devices. Some federal facilities, like MIT Lincoln Laboratory, already have such facilities that they make available. The private sector should make more of its facilities available to university and national laboratory researchers. Government investment is needed, because until a quantum industry develops further, there will not be enough demand for such facilities to be self-sustaining.

2. Actively foster collaboration and advance quantum ecosystems with bold objectives.

Ecosystems play a vital role in the development of advanced technologies. They typically incorporate a wide range of stakeholders, including academia, industry, investors, government, and nonprofit organizations. Healthy ecosystems promote knowledge transfers and foster innovation through both collaboration and competition. They play a vital role in nurturing talent and cross-pollinating ideas and act as an anchor for the most ambitious people entering the field. The United States is home to a number of embryonic quantum regional hubs that are typically dominated by universities and research institutes, including in Chicago, Illinois; Colorado; Massachusetts; and California.[23] There is also significant commercial activity, including companies building the equipment ("picks and shovels") for quantum computing. Some of those companies make components that are not exclusively quantum but can be adapted—such as microwave generators, lasers, optics, and electronics.

There are private-sector quantum computing efforts today that generate revenues but are not profit-making or self-sustaining. They largely survive through a mix of (1) venture capital funding, (2) government grants, (3) sub-scale commercial contracts, and/or (4) being part of a much larger technology company that is willing to invest in their long-term potential and to support the work until it becomes profitable. A profitable, scalable quantum computing business model is unlikely to emerge until quantum advantage is achieved and commercially viable use-cases are identified. This economic reality, coupled with the substantial national security and competitiveness imperative, means that federal investment is vital. Without it, much of the ecosystem would collapse. For that reason, today, some, though not all, leading private-sector firms receive federal funding.

While there is a strong quantum ecosystem foundation to build on in the United States, current initiatives need a clearer, bolder directives and goals—notably, the following:

- *Use funding to encourage bolder action and more collaboration.* Funding could be awarded in a way that encourages bolder action within the National Quantum Information Science Research Centers. Currently, research groups work largely independently. Larger-scale efforts often bump up against security concerns, which leads to smaller collaborations around specific scientific challenges rather than larger collaborations around broader challenges. Protecting security is obviously essential, but how security is maintained needs to be balanced against other goals. US security will ultimately be enhanced by having quantum computing advance to the point where a commercial industry can develop.
- *Leverage existing programs.* For example, use the US government's Small Business Innovation Research (SBIR) program to help promising quantum startups to accelerate their growth.
- *Incentivize collaboration for greater focus and scale.* As previously noted, more incentives for collaboration are needed. Existing consortia work well for road mapping and education around policies and regulations, but they lack the focus and scale to attempt, for example, building a next-generation quantum computer. The creation of shared facilities (or sharing of

existing facilities) that could provide access to devices or fabrication without revealing IP could be helpful.

- *Expand academic-industry collaboration.* There are many upsides to these collaborations, including expanding resources, targeting fundamental research toward industry goals, and developing talent. But given the importance and the sensitivity of the work, particularly around intellectual property (IP), it can be complicated to create highly effective partnerships. Leaders from both academia and industry should convene to discuss how to improve these collaborations and build on current models.

  A host of issues can complicate and slow down academic-industry partnerships. A potentially relevant model from the semiconductor industry is the Semiconductor Research Corporation. The work at Google on the "Willow" quantum chip is a good example of how collaboration between industry and academia can work.[24]

3. Accelerate workforce development.

Building a skilled workforce for an emergent quantum industry is vital and requires both developing domestic talent as well as cementing the United States as the top destination for talented people from around the world who want to research and work with quantum technologies.

Up to 2020, the quantum computing workforce was dominated by university-based physics researchers and professors. That is changing. Today many industry and academic professionals from fields such as electrical engineering, computer science, process engineering, and software development are shifting jobs or their focus to contribute to quantum computing advancements. Expertise in complementary metal–oxide–semiconductor (CMOS) technology, the dominant technology used in modern computer chips, as well as classical computing and chip fabrication are all needed to help move quantum forward. One does not need to be a quantum scientist to help drive the field forward.

Most important among the steps that can be taken to accelerate workforce development are the following:

- Develop the domestic talent pipeline as outlined in the Center for a New American Security report starting with the K–12 curriculum.[25] Increase awareness of quantum computing as a science/technology field among younger generations by demystifying and reducing the inhibition to engage with the topic. Expand existing educational programs to incorporate quantum computing, for example, by adding quantum computing programming skills to existing computer programming courses.
- Support the expansion of professional development in quantum computing. Existing programs, like MIT's xPRO courses, have proven highly successful in developing new talent, especially for those in industry who want to pivot to the field as well as those in government formulating policy.
- Attract and retain international quantum talent. US policies should reform student visa and high-skilled immigration policies to encourage workers with relevant skills to come to, and stay, in the United States. This could include the following:
    - Providing international students who earn advanced STEM degrees from US institutions a clear path to a green card. A first step would include passing the bipartisan Keep STEM Talent Act,[26] which would exempt non-US nationals with advanced degrees in a STEM field from direct limitations on the number of immigrant visas granted per year.
    - Updating H-1B rules that support the hiring of physicists and other STEM workers. The 2024 ruling by the US Citizenship and Immigration Services regarding the definition of "specialty occupation" for H-1B petitions in emerging fields was a positive development.[27]

- Allowing international students applying for an F-1 visa to indicate they would like to stay in the United States after graduation.[28] Start work visa investigations when foreign graduate students arrive, not when they graduate.
- Fast-tracking visa applications for researchers from countries that have signed bilateral quantum cooperation agreements with the United States, such as the United Kingdom, Japan, and Germany.[29]

4. Increase US quantum supply chain resilience.

While the scale-up of quantum computers may be some years away, experience shows that it can take time to reconfigure global supply chains, particularly for key quantum inputs like rare earths, helium-3, and high-performance classical computing chips. The focus today should be on building domestic supply chains as well as working with allies and partner countries where appropriate. The United States should make the quantum supply chain as much of a priority as others that have been deemed critical to the country. The federal government should begin the process of identifying gaps and weaknesses in the supply chain[30] and addressing them through incentives for domestic production (through tax policy or use of federal government authorities) and through alignment with international partners.[31]

Conclusion

To secure long-term leadership in quantum computing, the United States must adopt a sustained and ambitious approach. This requires resilient funding of foundational research consistently over decades; facilitating bold, cross-sector collaborations to build vibrant quantum ecosystems; investing in quantum workforce development to ensure a pipeline of skilled talent; and strengthening the resilience of the US quantum supply chain against geopolitical and technological disruptions. The global race to achieve quantum advantage and scale up this important technology is a long-term effort of scientific, economic, and strategic significance.

The authors would like to thank the following colleagues for their contributions to this chapter: Jeffery Grover, Research Scientist, Engineering Quantum Systems group, MIT; Elif Kiesow, Research Affiliate, MIT Initiative on the Digital Economy, and Celia Merzbacher, Executive Director of the Quantum Economic Development Consortium.

Notes

1. Quantum science describes the behavior of light and matter on the atomic and subatomic scale. It emerged from the study of the smallest objects in nature.

2. Hodan Omaar, "The U.S. Approach to Quantum Policy," Center for Data Innovation, October 10, 2023, https://www2.datainnovation.org/2023-us-quantum-policy.pdf.

3. Sam Howell, "The Quest for Qubits," Center for New American Security, May 28, 2024, https://www.cnas.org/publications/reports/the-quest-for-qubits.

4. In 2024 the National Institute of Standards and Technology announced a set of encryption tools that can be run on classical computers that are designed to withstand the attack of a quantum computer. This is an important development, but there is an enormous body of work needed to implement these post-quantum encryption standards into all existing digital infrastructure.

5. McKinsey and Company, "The Rise of Quantum Computing," McKinsey & Company, accessed July 28, 2025, https://www.mckinsey.com/featured-insights/the-rise-of-quantum-computing.

6. Howell, "Quest for Qubits."

7. "National Quantum Strategy," Quantum.gov, https://www.quantum.gov/strategy/.

8. National Quantum Initiative Act, H.R. 6227, 115th Cong. (2018), https://www.congress.gov/bill/115th-congress/house-bill/6227/text.

9. Civilian R&D funding includes National Science Foundation (NSF), Department of Energy (DoE), National Institute of Standards and Technology (NIST) (Department of Commerce), and National Aeronautics and Space Administration (NASA);

military R&D funding includes Defense Advanced Research Projects Agency (DARPA), Army Research Office, Air Force Office of Scientific Research (AFOSR), Office of Naval Research (ONR), Army Research Laboratory (ARL), Naval Research Laboratory (NRL), Air Force Research Laboratory (AFRL), and the Office of the Under Secretary of Defense for Research & Engineering; intelligence community funding includes Intelligence Advanced Research Projects Activity (IARPA), Laboratory for Physical Sciences (LPS) (National Security Agency [NSA]); Federally Funded Research and Development Centers (FFRDCs) working on quantum include the following: Argonne, Brookhaven, Lawrence Berkeley, Lawrence Livermore, Fermilab, Los Alamos, MIT Lincoln Lab, Oak Ridge, Pacific Northwest National Laboratory (PNNL), Sandia; Office of Science and Technology Policy (OSTP) and the National Quantum Initiative Advisory Committee (NQIAC) play coordinating roles across the various federal entities. National Science & Technology Council, Subcommittee on Quantum Information Science, Committee on Science, *National Quantum Initiative Supplement to the President's FY 2025 Budget* (December 2024), https://www.quantum.gov/wp-content/uploads/2024/12/NQI-Annual-Report-FY2025.pdf.

10. Quantum Sandbox for Near-Term Applications Act of 2025, S. 1344, 119th Cong. (2025), https://www.congress.gov/bill/119th-congress/senate-bill/1344/text.

11. National Quantum Initiative Reauthorization Act, H.R. 6213, 118th Cong. (2023), https://www.congress.gov/bill/118th-congress/house-bill/6213/text.

12. "Cantwell, Young, Durbin, Daines Introduce National Quantum Initiative Reauthorization Act," US Senate Committee on Commerce, Science, & Transportation, December 3, 2024, https://www.commerce.senate.gov/2024/12/cantwell-young-durbin-daines-introduce-national-quantum-initiative-reauthorization-act.

13. James Woodford, "Australia Places A$1 Billion Bet on Quantum Computing Firm PsiQuantum," *New Scientist*, April 30, 2024, https://www.newscientist.com/article/2428987-australia-places-a1-billion-bet-on-quantum-computing-firm-psiquantum/; Abby Miller, "Former U.S. Steel South Works Picked for Quantum Computer Site, with Multibillion-Dollar Investment," *Chicago Sun-Times*, July 25, 2024, https://chicago.suntimes.com/technology/2024/07/25/psiquantum-steel-south-works-quantum-computer-johnson-darpa.

14. It is difficult to obtain specific numbers on quantum investments from these large corporations as they are not required to report them. However, our internal analysis strongly suggests that US corporations lead total investment globally.

15. Elliott J. Mason III, "State of Quantum Industry Innovation—What Patents Tell Us," QED-C, December 11, 2024, https://quantumconsortium.org/blog/state-of-quantum-industry-innovation-what-patents-tell-us/.

16. Jonathan Ruane, Elif Kiesow, Johannes Galatsanos, Carl Dukatz, Edward Blomquist, and Prashant Shukla, "The Quantum Index Report 2025," MIT Initiative

on the Digital Economy, Massachusetts Institute of Technology, May 2025, https://qir.mit.edu/wp-content/uploads/2025/06/MIT-QIR-2025.pdf, 103; "Just Released: 2025 MIT Quantum Index Report," MIT Initiative on the Digital Economy, June 5, 2025, https://ide.mit.edu/insights/just-released-2025-mit-quantum-index-report/.

17. Materials refers to those that exhibit unique quantum properties like superconductors, topological insulators, graphene, and certain types of semiconductors, which can be used to create qubits for quantum computing.

18. See MIT xPRO, "Catalog," https://xpro.mit.edu/catalog/?topic=Technology%3AQuantum%20Computing.

19. American Physical Society, Office of Government Affairs, "Building America's STEM Workforce," January 2021, https://res.cloudinary.com/apsphysics/image/upload/v1681415466/Building-America-STEM-workforce-1_jtukoc.pdf; Jacinta Conrad, "Chips and Science Bill Isn't Enough. America Needs to Retain Its International Students," *APS News*, August 17, 2022, https://www.aps.org/apsnews/2022/08/chips-and-science-bill.

20. Howell, "Quest for Qubits," lists the following known supply chain inputs: Materials: helium-3 gas, silicon-23, copper, aluminum, and gold; components and subassemblies: high-performance cryocoolers, pumps, valves, compressors, power supplies, RF generators, superconducting wiring assemblies, dilution fridge components, fiber and coaxial cables, low-noise lasers, and key manufacturing equipment for quantum and classical chip manufacturing and testing.

21. See "Supply and Demand of Helium-3," US Department of Energy, https://science.osti.gov/np/Research/IDPRA/3He-Fact-Sheet.

22. "About IMEC," IMEC, https://www.imec-int.com/en/about-us.

23. Matt Swayne, "Tomorrow's Quantum Hotbeds? 7 U.S. Cities That Could Incubate the Next Great Quantum Technology Ecosystem," *Quantum Insider*, April 11, 2024, https://thequantuminsider.com/2024/03/04/tomorrows-quantum-hotbeds-7-u-s-cities-that-could-incubate-the-next-great-quantum-technology-ecosystem/.

24. Google Quantum AI and collaborators, "Quantum Error Correction Below the Surface Code Threshold," *Nature* 638 (2025): 920–926, https://doi.org/10.1038/s41586-024-08449-y.

25. Howell, "Quest for Qubits."

26. Keep STEM Talent Act of 2023, S. 2384, 118th Cong. (2023), https://www.congress.gov/bill/118th-congress/senate-bill/2384/text.

27. Leslie Dellon, "H-1B Modernization Rule Provides Some Comfort but Also Raises Concerns," *Immigration Impact*, American Immigration Council, December

20, 2024, https://immigrationimpact.com/2024/12/20/h1b-modernization-rule-provides-some-comfort-but-also-raises-concerns.

28. American Physical Society, "Building America's STEM Workforce."

29. "Enhancing Competitiveness," Quantum.gov, https://www.quantum.gov/competitiveness/.

30. The new Supply Chain Center at the Department of Commerce could potentially be a partner in this effort. See "The Department of Commerce's First-of-Its-Kind Supply Chain Center," International Trade Administration, https://www.trade.gov/supply-chain-center.

31. Howell, "Quest for Qubits," goes into more detail about building supply chain resilience in the quantum industry.

5
Drones: Advancing US Innovation in Autonomy and Opportunities for Reindustrialization

Fiona E. Murray

Strategic Importance

Drones, a catch-all term for uncrewed systems, represent a transformative technology platform—especially now that artificial intelligence (AI) renders them fully autonomous—with profound implications for national security and global competitiveness. The term "drone" was first coined in the 1930s to describe the Queen Bee, a British radio-controlled biplane whose buzzing evoked swarms of bees.[1] While most discussions of modern drones have focused on uncrewed aerial systems (UAS), today's drones also include uncrewed ground systems (UGS) and uncrewed maritime systems (UMS). With the growing number of acronyms, they are often referred to collectively as "uncrewed X systems" (UxS).

Modern drones are typically categorized by size and range, with the most well defined being those for UAS where NATO has established a range. These established ranges are Class I: micro and mini drones (e.g., traditional hobbyist drones, such as the Chinese-produced DJI Mavic and US-produced Skydio X2); Class II: medium-sized tactical drones (such as Turkey's Bayraktar TB2 drone); and Class III: medium- and high-altitude drones (such as General Atomic's Predator and Reaper drones).[2]

Today, advances in semiconductors, sensors, energy storage, and propulsion enable drones to operate in the air, space, land, and sea, with AI now allowing full autonomy. At their core, drones have sophisticated software systems capable of autonomous navigation and control using increasingly powerful algorithms enhanced by improvements in computing power and AI integration. These software systems depend on a wide array of sensors—from cameras and thermal imagers to radar, lidar (light detection and ranging), and GPS receivers—integrated into physical platforms composed of rigid airframes, propulsion systems, energy- and power-management systems, and auxiliary subsystems such as avionics and communications.

Beyond the systems themselves, an important factor in maintaining competitive advantage in drones is the ability to manufacture them at scale and pace, particularly in the event of military conflict where scaled production is likely to be essential to deploying drones over a sustained period of time. Key expertise that is required for drone production includes advanced manufacturing technologies and robotic assembly plus advances in a range of sensors, as well as both high-end and low-cost semiconductors. It also requires critical supply chains that provide components, some of which include rare-earth elements and critical minerals (e.g., for magnets or semiconductors).

Because each of these technologies is covered at length in other sections of this book, we will only briefly address them here in the context of drones and autonomy. Policy recommendations required to bring the supply chains for drones back to the United States—including semiconductors, motors, batteries, and other electronics—are similar to those required to bring components for other applications back to the United States. That said, we note that the particular needs for military preparedness and production at scale in the event of a national emergency or war-footing present their own policy difficulties in terms of the need for spare capacity, supply chain resilience, potential for stockpiling, and other logistical issues.

Drones, regardless of where they are produced, have a wide variety of military and civilian applications. Until recent military conflicts, drones were bifurcated between large military (Class II and III) platforms that provided intelligence, surveillance, and reconnaissance (ISR) or strike capabilities, and small civilian (Class I) platforms that provided capabilities such as those for agriculture and delivery. Indeed, in many ways, Class I platform drones were predominantly considered to be for hobbyists and certainly not considered to be a dual-use (e.g., military as well as civilian) platform.

However, the recent conflicts in the Caucasus, Ukraine, and the Middle East have driven a convergence in the military and civilian drone markets and the increasing use of Class I drones as a dual-use solution for military as well as civilian applications. Modern warfare has made extensive use of small consumer drones (especially in Ukraine), and has demonstrated that they can be pressed into military service (with only limited adaptations) for use as a weapon against costly traditional military equipment.[3] These conflicts have also resulted in a buildup of significant assembly and production capabilities; drone production that started in basements and bedrooms is now occurring in larger facilities, some with the capacity to produce over five million drones per year.[4]

Lessons from conflict have shifted attitudes toward the role of drones across the entire UxS spectrum with the recognition of the potential power of "precise mass"—that is, many cheap drones coupled together with accurate autonomous software and sensor systems to allow for precise surveillance and targeting.[5] While not a substitute for sophisticated aircraft systems such as the F-35 or B-2, these drones and the AI-enabled autonomous systems that support them are a growing part of military planning and procurement for the United States and its NATO allies and partners.[6]

Current Landscape

The landscape for drones and autonomy has evolved over the last two decades, and today's locus of innovation in ecosystems and in the industrial base (for production at scale) can best be understood by comparing small commercial drones (which are increasingly crossing over into military applications) with traditional military drones.

The commercial drone market is dominated by Chinese company DJI (Da-Jiang Innovations), with around 60 percent of the global market and 80 percent of the US market—even as US law bars its military from buying Chinese drones.[7] In line with Beijing's "Made in China 2025" (MIC25) initiative, DJI has become a market leader, and it remains a pioneer and innovator in the industry. For example, it was among the first to integrate technologies such as gimbaled high-definition cameras, lidar technology, and GPS receivers. Collectively, these technologies augmented the capabilities of human drone pilots and set the stage for autonomous operations. The company had been founded in 2005 by Frank Wang while at the Hong Kong University of Science and Technology. In 2013, it announced its first ready-to-fly consumer drone—Phantom 1—a turning point toward mainstream consumer drones.

Reductions in costs along with expanding functionality (at low cost) popularized drone use in applications from infrastructure inspection to search and rescue and law enforcement, with the global market estimated to be around $40 billion. Growth in the drone industry is currently estimated at around 10 percent per annum, especially in the fields of last-mile delivery and agriculture, driven by innovations in improved battery life and autonomous flying capabilities.[8] As of 2025, the United States has over 850,000 drones registered (the majority of which are from DJI), and over 250,000 certified drone pilots.[9]

Commercial drone production is also dominated by China. In 2015, DJI's Shenzhen-based production—which is deeply embedded within the extensive global electronics industry supply chain and the multi-industry manufacturing base—produced at least one million drones. In contrast, the entire US drone manufacturing sector—then comprising approximately 500 companies—produces fewer than 100,000 units annually.[10] Commercial drone assembly and supply chains in China also benefit from decades of industrial scale and specialization, driving down costs and increasing opportunities for innovation. That said, such Chinese dominance was not inevitable: many key technologies essential for the entire unmanned autonomous system had their origins elsewhere. For example, our research has shown that while the essential components in today's Class I drones were invented in the United States, Europe, and South Korea, they are now typically only available for assembly at scale in China (see figure 5.1). With limited supply chain accessibility in the United States and elsewhere, much of the current innovation now goes hand in hand with production in China.[11]

The full supply chain back to key material inputs is now also dominated by China due to its control of some critical materials for key components, such as rare earths for magnets (and the underlying rare earth metal processing) that form an essential part of brushless motors, lithium for the specialization lithium polymer batteries, and the low-cost electronics in drones.[12]

In the military arena, key national advantages with respect to drones are built upon at least three capabilities: technological expertise (in a range of areas), production capacity (including access to supply chains), and operational experience (which allows for in-field innovation and experimentation). As a result of decades-long investment in innovation, the United States is recognized as a world leader in high-end military drone technologies.[13] This expertise is dominated by traditional defense companies (so-called prime contractors) such as General Atomics, which produces Predator and Reaper drones for ISR, and Lockheed Martin, which produces stealth drones as well as those designed to operate alongside

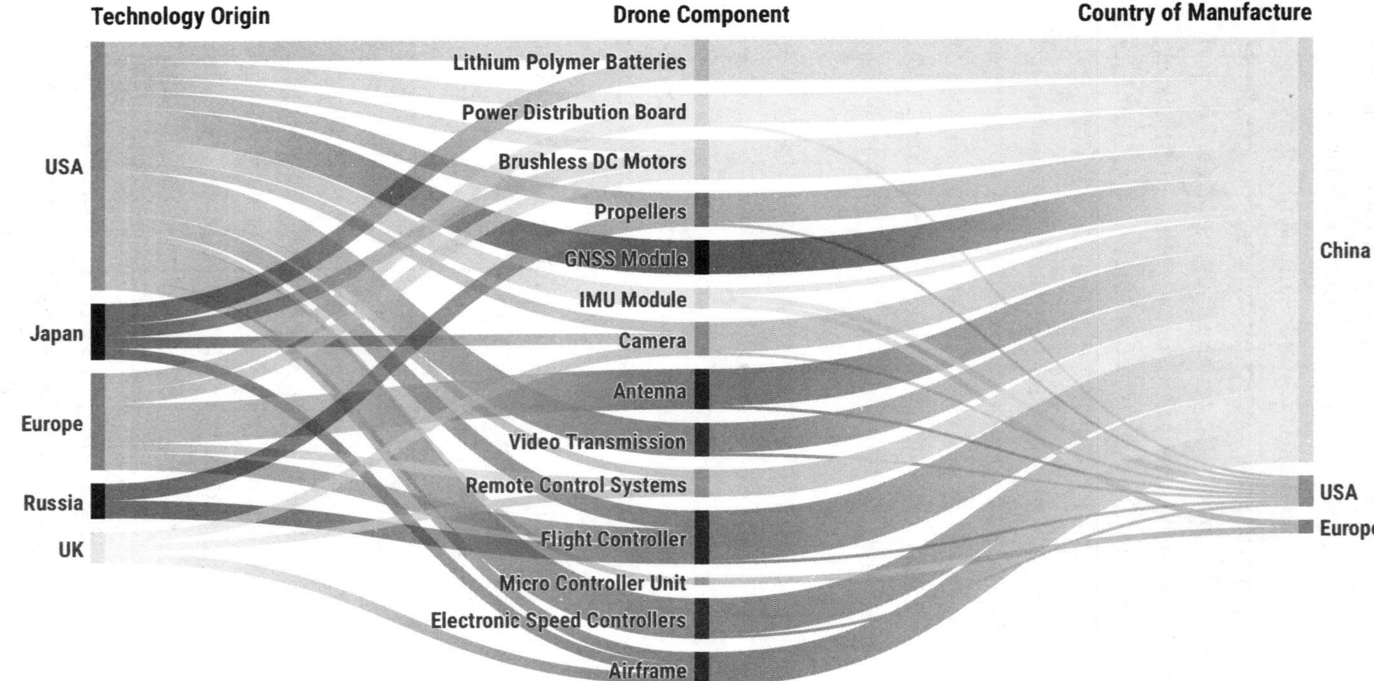

Figure 5.1
A diagram of the origin of drone component technologies (left) versus where they are manufactured today (right). Despite innovations emerging from the United States, Europe, Japan, South Korea, Russia, and the United Kingdom, China dominates large-scale assembly of almost all drone components. Source: Edlyn V. Levine and Fiona Murray, "How the US and Its Allies Can Rebuild Economic Security," *MIT Technology Review*, July 30, 2024, https://www.technologyreview.com/2024/07/30/1095439/usa-economic-security-competitiveness/.

its crewed aircraft platforms. Outside the United States, Israel has been a leading nation in the design and production of military drone technologies since the mid-1970s. The country's production is largely based on domestic and allied supply chains.[14] Much as in other arenas of military technology, European defense corporations have also provided some limited drone-based capabilities.[15] And while less is known about the specifics, the Aviation Industry Corporation of China produces a range of autonomous systems.

Beyond prime contractors, the military drone/autonomy industry in the United States and beyond has been shaped by new startup ventures that have largely entered with expertise in software and AI. These enable drones to become uncrewed systems (rather than simply vehicles) and for systems to be operated in swarms in complex and contested environments. The rise of AI/autonomy startups in defense has been driven by civilian uses related to resilience (e.g., driverless vehicles, infrastructure inspection, and air traffic). But more recently, pure defense US startups in autonomy/AI have emerged: Shield.ai, founded in 2015, has raised over $1.3 billion, and Anduril, founded in 2017, has raised over $6 billion by 2025. While both started with a focus on AI in military systems, they have integrated with hardware. Moreover, while open-source platforms have been essential for drone software control, startup solutions have focused on integrating sensor data for intelligence gathering, self-navigation, and swarming capabilities. Taken together, military and civilian UxS ventures in the United States and Europe have received over $3.5 billion in venture funding by 2025 (with 80 percent coming from the United States), up from $450 million in 2021.

Given its recent battlefield experience and the necessity for innovation, Ukraine also has gained significant expertise for small drones in every part of the drone/autonomy stack and across all three domains. Their proximity to the front lines has led to the development of a robust and vibrant innovation ecosystem (including entrepreneurs, risk capital, universities, government, and large corporations). Under wartime conditions, Ukrainian drone entrepreneurs, backed by venture capital (including the fund D3, formed after the start of the conflict to support Ukrainian entrepreneurs), worked closely with frontline forces through rapid experimentation cycles, eventually bringing the government and its corporate primes along. Indeed, a fourth branch of Ukraine's military—the Unmanned Systems Forces—now serves as a demanding customer for these new companies, and for novel products with production at scale. Given the conflict, they have also built production expertise (from sourcing parts through global supply chains), although as of 2025, export of these drones to markets outside Ukraine is prohibited by its government. Further, especially since the start of the war in Ukraine, a growing willingness in Europe to invest in dual-use and even some defense-focused ventures has seen the rise of drones and autonomy startups receiving significant venture capital. For example, Helsing in Germany is developing and manufacturing military drones as well as related software.

Also notable was the entry of Turkish company Baykar, with its medium-sized Bayraktar drones in the mid-2000s. The company, founded by the Bayraktar family in 1984, diversified from automotive parts production into military drones in the early 2000s, with the return of the founder's son to become the CTO after his studies at MIT. With his brother (in the CEO role), the firm produced drones at increasing scale with domestic supply chains, taking advantage of large-scale automotive and consumer white-goods production in the country. Exports of Bayraktar drones have grown significantly in the past several years, ending up with forces in both Azerbaijan and Ukraine. Given shifting global coalitions, expertise in different types of military drones has become a source of comparative advantage that drives both national security and economic growth.

Gaps and Opportunities

There are several areas where the United States could strengthen its drone innovation and industrial production capacities.

Supporting Innovation in AI-Based Autonomy and Sensors for Drones

The United States has significant expertise in autonomy and AI, including extensive compute infrastructure and a wealth of talent and expertise in agentic AI and autonomy-based applications in civilian sectors such as manufacturing robotics, autonomous driving, and delivery services. That said, China has a strong position in AI publications and patents with demonstrable work in open-source AI/agent-based platforms. To build competitive advantage, the United States must ensure continued commitment to research and training in AI as well as activities along the technology readiness levels, including in the regulatory context for testing and evaluation (see below).

In the hardware that is essential for effective AI/autonomy, the United States continues to have significant advantages. This is especially true in key sensing technologies such as cameras and lidar (which uses laser pulses to create detailed 3D maps for navigation, obstacle avoidance, and surveying). The United States also leads in ultrasonic sensors (using sound waves for short-range obstacle detection). Advanced radar systems, long an area of US leadership, also continue to be important to detect and track multiple targets simultaneously in various weather conditions. The United States has a long commitment to fundamental research in this area. To acquire this lead, the United States has made substantial investments in sensor technologies through the Department of Defense's Defense Advanced Research Projects Agency (DARPA) and national labs, such as MIT Lincoln Labs.

However, China is developing significant advantages in specific areas, especially sensors in the maritime domain. For example, they are making advances in magnetic detection technology, which makes it more straightforward for them to detect US submarines, and uniform messaging systems, with their coherent population-trapping atomic magnetometer. Similar advances in UAS and counter-UAS systems are also being expedited in China, suggesting that US leadership cannot be assumed.[16]

Production of Drones and Their Critical Components

One of the largest gaps in US leadership in drones and autonomous systems is the lack of supply chain concentration like the one found in Shenzhen, China. The latter's "one-hour supply chain" is made up of thousands of vendors for brushless motors, electronic-speed controllers, gimbals, lithium-ion cells, cameras, and micro-electromechanical sensors. Originally built up for smartphones and consumer electronics, this cluster now provides a significant low-cost and sophisticated industrial base to support a growing innovation ecosystem. DJI itself relocated within China to Shenzhen to take advantage of this momentum. Even with significant subsidies, it would be challenging to build out drone manufacturing in the United States at scale without similarly scaling advanced manufacturing, critical minerals, and semiconductors.

That said, the recent years of war have helped Ukraine become a drone powerhouse, and the United States can benefit from partnering with it, perhaps importing some of its battle-tested drones when its export ban is lifted. At home, the United States has an opportunity to rebuild its production at scale. For example, Anduril is building "Arsenal-1" in Ohio, a hyperscale manufacturing facility to produce autonomous weapons systems, including aerial and maritime drones at scale. However, until supply chains mature domestically, US-based drone manufacturing will remain vulnerable to disruptions and sanctions imposed by China. This vulnerability was clearly illustrated in late 2024 when Skydio faced severe supply chain disruptions after China sanctioned the company for selling drones to Taiwan. The sanctions explicitly prohibited Chinese companies from supplying Skydio with critical components, most notably lithium-ion batteries, significantly impacting its ability to manufacture drones at scale.

China intensified these pressures by implementing broader export bans on gallium, germanium, and antimony—all critical minerals essential for drone subsystems such as motor drivers, lidar systems, and thermal imaging sensors. These minerals are indispensable due to their unique properties: gallium is pivotal for power electronics and semiconductors; germanium is essential for infrared optics used in advanced imaging systems; and antimony plays a critical role in semiconductor manufacturing and battery technology. Until it establishes robust, diversified, and domestic or allied sources of these critical components and materials, the US drone industry will remain at substantial risk of geopolitical disruptions.

Fragmented Regulatory and Testing Environment

Despite strong US leadership in AI and autonomy, companies face major barriers in testing and evaluating autonomous systems due to fragmented and unclear regulatory environments. On land, the lack of clarity and significant complexity in the civilian context are placing limits on autonomy in key civilian sectors such as the automotive industry, where China is in a leading position in the deployment of semi- and fully autonomous systems on the roads. In the air, the absence of national testbeds and streamlined rules for beyond visual line of sight (BVLOS) operations impedes the translation of advanced autonomy software into operational military or civilian drone platforms in civilian airspace. In the marine environment, similar complexity exists and is limiting current efforts at US leadership.

Demand-Side Pull

Although the United States excels at making and buying some of the most sophisticated drone/autonomous systems for military purposes, through its annual National Defense Authorization Act, the rise of mass-produced drones and the convergence of civilian and military solutions require a new approach to procurement. Both civilian demand (enabled by clear testing and regulatory frameworks) and military demand (for small drones supplied at scale) are essential. With clear demand-side pull comes contracts and annual recurring revenues that spur investors into a sector, reduce uncertainty, lower the cost of capital, and allow for expanded production. Without clear demand-side approaches, the United States will struggle to produce commercial-like products at a predictable cadence. On the other hand, countries such as China, Turkey, and even Ukraine are placing orders in the multi-year, multi-thousand-unit scale in both the civilian and military markets that seed demand and enable manufacturers to access reliable financing at every stage of the product (or system) life cycle. This in turn drives the scale and scope of production and the coupling of vibrant innovation ecosystems to a robust industrial base and its supply chains.

Recommendations

For the United States to become a more resilient developer, producer, and exporter of drones and their autonomous systems, it will need to maintain a commitment to AI/autonomy research and development, while significantly overhauling its policy approach to testing and evaluating autonomous systems in all domains. In addition, the United States must recommit to integrated supply- and demand-side policies around the drone/autonomy industrial base, including for advanced manufacturing capacity and supply chains. This is particularly crucial with regard to subsystems or components such as semiconductors, electronics (such as high-energy/power-density motors) and energy storage systems. It is also important to consider the development of more resilient supply chains for these elements (especially when it comes to their critical mineral requirements). Some of these topics are covered in-depth across the chapters that discuss critical minerals, semiconductors, and advanced manufacturing in this volume, so we will only cover recommendations in this section that are unique to the dual-use nature of drones/autonomy or that go beyond what is recommended in other sections.

1. **Integrate and synthesize data and insights from across a range of domains.**

- The United States has inadequate investment in testing environments (known as "innovation ranges," much akin to a firing range for testing novel weapons) that allow early-stage autonomy startups to test and evaluate at pace. This leads to gaps and frictions in private-sector investment. Specifically, the government should do the following:
  - Develop testing areas and testbeds that match the needs and challenges of various domains, including air, land, and ocean testing and evaluation.
  - Emphasize the provision of autonomous platforms in the ocean within a framework of Coast Guard regulation. This could draw upon the US Coast Guard's pilot program under which SpaceX's unmanned barge operates to recover spacecraft.
- Launch a clear set of guidelines and access rules for drone testbed facilities, making those within the US National Lab infrastructure (including MIT Lincoln Labs) accessible to appropriate participants. Such a program should do the following:
  - Make clear the regulatory environment around testing, particularly for airborne systems, given the difficulty of testing

drones in civilian air space under current regulations; likewise for maritime environments.

- Consider using existing national security waivers to limit the application of civilian regulations, and to grant these liberally to companies that will test and evaluate autonomous systems for national security applications.
- Validate specific platform or product testing protocols such that, once completed, products can be certified as meeting certain specifications.

- Support programs for the rapid demonstration of specific autonomous capabilities in particular military or civilian contexts to allow for greater understanding of requirements and to offer clear demand signals. For example, the US government should do the following:
  - Establish flagship software programs (akin to major hardware platforms like the F-35) and fund them commensurately to signal strategic priority and drive consistent investment in cutting-edge autonomy and AI.
  - Expand, with allies, upon activities such as the NATO Baltic-X task force, which provided an opportunity for startups, scale-ups, and prime contractors to demonstrate and integrate their drone capabilities in the maritime context. Likewise, the government must encourage demonstrations that illustrate the integration of military and civilian drone-based activity in the same airspace.[17]
  - Undertake projects that allow for different types of drones to be connected to a common platform to allow for much-needed variety in drone hardware and drone capabilities. Germany's Taskforce Drone has undertaken such an analysis, learning from Ukraine the need for platforms that adapt to in-field iteration and learning cycles.[18]

- As part of its commitment to testing, evaluation, and wider deployment, policies that promote open but protected data sharing in training exercises and simulations (e.g., federated learning models) and greater regulatory clarity would be valuable. Specifically:
    - Interagency discussions should be explored to enable the expansion of the civilian markets for drones.
    - To achieve sufficient scale and to allow for effective simulation, a well-funded national uncrewed aircraft traffic management system with data exchange standards would expand civilian applications.
    - To further unlock these civilian markets, and in line with the June 2025 Executive Orders "Restoring American Airspace Sovereignty" and "Unleashing American Drone Dominance," the Federal Aviation Administration (FAA) (working in collaboration with NASA for low earth orbit and the US Department of Defense) should issue its final BVLOS rules and pilot programs. This would expand civilian use of drones in a range of applications, boosting domestic demand for US civilian and dual-use platforms while ensuring control over US airspace, protecting critical infrastructure, and bolstering public safety.

2. Establish national capability for advanced drone manufacturing and related supply-chain production.

The United States needs to establish national capability for advanced drone manufacturing and related supply chain production by building capacity within regional innovative industrial-base hubs around the country where expertise in both civilian and military products could be combined or shared in times of crisis.

The ability to produce drones at scale and on demand is especially important in times of crisis. However, stockpiling approaches are unlikely to be successful due to the ever-changing nature of drone technology and the need to update at the pace of conflict. Future government-sponsored capacity should therefore be focused on supporting regionally distributed facilities that can also serve other adjacent and complementary activities. These facilities might also serve as a setting for development, testing, and training of more advanced in-country drone assembly and supply chains. Specific actions might include the following:

- Support the development of US regional hubs with a dedicated "drone foundry" that would provide necessary space and support for small-scale drone manufacturing and supply-chain component development. As such it would support and coordinate the emergence of a broader drone manufacturing ecosystem that might also include education, workforce development, training facilities, and production capacity. These hubs should have incentives to prioritize innovation in drone development and in key supply chain choke points. In order for these foundries to be successful, the following conditions would be crucial:
    - Ensure that sustained and predictable funding is available to maintain the infrastructure and skilled workforce necessary for production, ensuring economic stability, and preventing workforce reductions.[19]
    - Design hubs to work with key US partners and allies that have particular complementary capabilities in drone production and supply chains so that the United States can benefit from this innovation and industrial base expertise. The recent Artemis program led by the Defense Innovation Unit, which includes two Ukrainian companies in partnership with US participants, is a case in point.[20]
    - Support the development of partnerships between US regional manufacturing hubs and specific hubs of expertise and scale in allied and partner nations. For example,

Ukraine, with capacity of over five million drones per annum, could provide essential support of US capacity needs while securing the drone supply chain of the United States and its allies against foreign control or exploitation.

- Continue prior technical support for expanded Ukrainian production with closer ties to US centers of manufacturing to enable know-how transfer and training.[21]

3. Require supply chain risk management of autonomous systems, including hardware and components.

Particularly in a defense context, it is essential that the government and its suppliers can map, monitor, and mitigate vendor and country-of-origin risks. This will help not only to reduce supply chain vulnerabilities but also to guide supply-side incentives for strengthening activities in drones/autonomy. A priority should be ensuring government and/or industry requirements to use traceability and provenance approaches that are low-cost and can be widely adopted. As noted in chapter 1, on critical minerals, similar approaches would also support more careful policies for critical materials transparency when combined with specific policy mandates (such as stockpiling of certain defense-relevant critical materials). Specifically:

- The United States could require that, for military drones, imported critical minerals and the products that use them are processed within a specific set of nations. This would create a wider market for domestic and partner-nation mineral producers.
- The US Department of Defense should work alongside international bodies including NATO, the G7, and the European Defence Industrial Strategy for the development of trusted global supply chains in defense-related critical materials. A recent starting point is the NATO critical materials analysis for defense equipment, which highlights key dependencies. The

inclusion of drones in the next iteration of this analysis would be particularly valuable.

4. Implement demand-side strategies that encourage innovation and scaling in all aspects of drones/autonomy in both defense and civilian markets.

These strategies should also include civilian markets, especially those with government buyers, and they should promote efficiencies in the supply chain in order to drive economies of scale. Key strategies in this area include the following:

- Federal procurement and other demand-side policies should be aligned around drone/autonomy software capabilities rather than focusing on procuring specific products. Specifically:
  - Purchase autonomy software as a standalone product (much as the civilian sector buys software through licenses) and not just as part of a hardware package, thus allowing companies to license software while retaining control of intellectual property (IP).
  - Enable faster export approvals for advanced autonomy software systems, especially in situations involving conflict where large-scale deployment of drones with sophisticated software systems may be essential to allied and partner nations to allow for market expansion, and to reestablish global opportunities for US-developed capabilities in the face of growing international competition, as well as market expansion to drive production and supply chains at scale.
  - Ensure that commercially developed solutions are considered for software as well as hardware from all parts of the innovation ecosystem, including startups and established firms, and not just bespoke systems.
- Design US and allied procurement for drone/autonomy hardware that encourages innovation and capacity building for US

manufacturing and assembly, such as contracts that buy capacity and capabilities rather than just platforms.

- Expand the US Department of Defense's Replicator program. This one-time initiative led by the Defense Innovation Unit to apply new procurement methods to buy drones should be extended into a five-year rolling advance market commitment covering at least 50,000 units.
- Use procurement and other demand-side policies to encourage targeted innovation in key components where significant dependencies have arisen, notably in battery pack design, where overreliance on China has significantly weakened supply chain resilience.
- Look further down the supply chain into critical materials, with expansion of market commitments into mining and processing by the United States and its allies.[22]

- The government should define "trusted drones" as those with supply chain security aligned with national security interests to mitigate concerns about drones sourced from adversarial nations. Demand-side policies could also be harmonized across federal departments in the United States to enable greater economies of scale and scope for the United States, its allies, and partners in this strategic sector. Specifically:
  - Departments beyond the Department of Defense should be required to purchase from existing DoD suppliers. For example, expand the DoD Blue UAS list—a curated list of drones approved for government use on the basis of cybersecurity, supply chain integrity, and compliance with the National Defense Authorization Act—thereby reducing the barrier for small manufacturers to sell to the federal government.[23]
  - Expand government demand for drones beyond the Department of Defense. For example, the US Coast Guard could benefit from strategic procurement of autonomous systems

that strengthen maritime domain awareness. In addition, NOAA's missions could be enabled by autonomous maritime systems and could support wider United States economic security goals.

- ◦ Department of Homeland Security missions could be supported by a range of autonomous systems, particularly in the air and for forest fire detection and response, driving demand for trusted drones.

5. Implement financial incentive strategies that encourage manufacturing scale-up of drones and their critical components in the United States and in allied and partner nations.

- Establish a targeted investment approach, including tax credits, loan guarantees, and matching grants for trusted drone airframes, battery packs, motors, and navigation chips integrated into a national drone and autonomy manufacturing initiative. Mechanisms might include the following:
    - ◦ Investment tax credits to incentivize startup ventures and small- and medium-sized enterprises to invest in developing drone production or key drone components. Tax credits could also be structured to provide incentives to build domestic manufacturing, including via the DoD Office of Strategic Capital.
    - ◦ Federal programs should provide low-interest, long-term loans to support manufacturing equipment development and supporting software/automation, and to build out capacity for drone assembly and component development. The EXIM Bank's Make More in America program is a potential source of funding as are the Bank's resources more broadly, alongside a specific focus on prioritizing the export of US-manufactured civil uncrewed aircraft systems.[24]
    - ◦ Encourage US investment, potentially through entities like the International Development Corporation, in dual use

drone/autonomy capabilities and production capacity in allied nations, including in Ukraine, to ensure that opportunities to tap into lower-cost manufacturing and assembly close to real-time innovation expertise are integrated into US priorities.

Conclusion

A US-led drone economy will require a commitment to a strong industrial base to allow for scaled production, supply chain security, and robust friendshoring. These conditions will allow a US drone economy to thrive and to serve as an "arsenal of democracy" unfettered by dependence on adversaries in the event of war. At the same time, drones and autonomous systems must continue to draw from innovation ecosystems in the United States and its allies and partners in order to support novel hardware and software capabilities. Finally, we must continue with the rapid experimentation and adaptation that have characterized drone innovation cycles throughout the war in Ukraine.

The author would like to thank the following colleagues for their contribution to this chapter: Phil Budden, Senior Lecturer, MIT Sloan School of Management; Rory Burke, MBA 2025 MIT Sloan School of Management; and Gene R. Keselman, Lecturer, MIT Sloan School of Management, Executive Director, MIT Mission Innovation Experimental.

Notes

1. John Buckley, *Air Power in the Age of Total War* (Bloomington: Indiana University Press, 1999), https://iupress.org/9780253213242/air-power-in-the-age-of-total-war/.

2. Andre Haider, *A Comprehensive Approach to Countering Unmanned Aircraft Systems* (Joint Air Power Competence Centre, 2021), https://www.japcc.org/chapters/c-uas-introduction.

3. Daryna Ostafiichuck et al., *Ukraine Drone Ecosystem Analysis: Building & Leveraging Ukraine's Drone Capabilities in Conflict & Beyond* (MIT Sloan, 2025), https://murray-lab.org/wp-content/uploads/2025/03/UkraineDroneEcosystem-Analysis_2025.pdf.

4. John Thornhill, "Ukraine Is Winning the Drone Start-Up War," *Financial Times*, April 10, 2025, https://www.ft.com/content/2c7d3c96-f6a8-4afa-bd88-5ea2fa39f5c1.

5. "Gremlins," Defense Advanced Research Projects Agency (DARPA), accessed July 9, 2025, https://www.darpa.mil/research/programs/gremlins.

6. Michael C. Horowitz, Lauren A. Kahn, and Joshua A. Schwartz, "What Drones Can—And Cannot—Do on the Battlefield: The Pentagon Should Learn from Israel and Ukraine," *Foreign Affairs*, July 4, 2025, https://www.foreignaffairs.com/united-states/what-drones-can-and-cannot-do-battlefield; Laura Gozzi, "How Ukraine Carried Out Daring 'Spider Web' Attack on Russian Bombers," *BBC News*, June 2, 2025, https://www.bbc.com/news/articles/cq69qnvj6nlo.

7. Farah Stockman, "Drones Are Key to Winning Wars Now. The U.S. Makes Hardly Any," *New York Times*, July 13, 2025, https://www.nytimes.com/2025/07/13/business/drones-us-military-manufacturing-lags.html?smid=nytcore-ios-share&referringSource=articleShare.

8. "Drone Market Analysis, Size, Share and Growth Trends 2023–2032," Market.us, accessed July 9, 2025, https://market.us/report/drone-market/.

9. "Top 10 Commercial Uses for Drones," Inspired Flight, April 13, 2023, https://www.inspiredflight.com/news/top-10-commercial-uses-for-drones.php.

10. Stockman, "Drones Are Key to Winning Wars Now."

11. Edlyn Levine and Fiona Murray, "How the U.S. and Its Allies Can Rebuild Economic Security," *MIT Technology Review*, July 30, 2024, https://www.technologyreview.com/2024/07/30/1095439/usa-economic-security-competitiveness/

12. Keith Bradsher, "U.S. Dependence on China for Rare Earth Magnets Is Causing Shortages," *New York Times*, June 2, 2025, https://www.nytimes.com/2025/06/02/business/china-rare-earths-united-states-supplies.html.

13. Mert Durgut, "The Race for UAV Supremacy: Which Country Leads the Pack?," Aviationfile, November 22, 2023, https://www.aviationfile.com/the-race-for-uav-supremacy-which-country-leads-the-pack/.

14. Uri Sadot, "A Perspective on Israel," Center for a New American Security, May 1, 2016, https://drones.cnas.org/wp-content/uploads/2016/05/A-Perspective-on-Israel-Proliferated-Drones.pdf.

15. *Fortune Business Insights*, "10 Best Military Drone Manufacturers in the World, 2021," July 27, 2021, https://www.fortunebusinessinsights.com/blog/10-best-military-drone-manufacturers-in-the-world-10590#.

16. Tye Graham and Peter Singer, "China's Counter-UAV Efforts Reveal More Than Technological Advancement," Defense One, May 2, 2025, https://www.defenseone.com/technology/2025/05/chinas-counter-uav-efforts-reveal-more-technological-advancement/405031/.

17. NLR, "German Armed Forces Drone Flies Through European Upper Airspace for the Time," February 6, 2025, https://www.nlr.org/newsroom/nieuws/german-armed-forces-drone-flies-through-european-upper-airspace-for-the-time/.

18. Matthias Monroy, "Lessons Learnt from the War in Ukraine: German Armed Forces to Catch Up on Drones," digit.site36.net, July 16, 2024, https://digit.site36.net/2024/07/16/lessons-learnt-from-the-war-in-ukraine-german-armed-forces-to-catch-up-on-drones/ (originally published in German as "Bundeswehr soll Nachholbedarf zu Drohnen stillen," nd.aktuell, https://www.nd-aktuell.de/artikel/1183765.task-force-drohnen-bundeswehr-soll-nachholbedarf-zu-drohnen-stillen.html).

19. "AUVSI's 2025 Hill Day: Advocating for the Future of Autonomy on Capitol Hill," Association for Uncrewed Vehicle Systems International (AUVSI), accessed July 9, 2025, https://www.auvsi.org/auvsis-2025-hill-day-advocating-for-the-future-of-autonomy-on-capitol-hill.

20. Joe Saballa, "US Selects Four Firms, Including Ukrainian Partners, for Artemis One-Way Drone Program," *Defense Post*, March 17, 2025, https://thedefensepost.com/2025/03/17/us-artemis-drone-program/.

21. Julian Barnes, "U.S. Reveals Once-Secret Support for Ukraine's Drone Industry," *New York Times*, January 17, 2025, https://www.nytimes.com/2025/01/17/us/politics/ukraine-drones-biden-support.html.

22. "Clean Investment Monitor," Rhodium Group and MIT's Center for Energy and Environmental Policy Research, accessed July 9, 2025, https://www.cleaninvestmentmonitor.org/.

23. Valerie Heitshusen and Brendan W. McGarry, *Defense Primer: The NDAA Process* (Congressional Research Service, January 6, 2025), accessed July 9, 2025, https://www.congress.gov/crs-product/IF10515.

24. Executive Order 14307, "Unleashing American Drone Dominance," 90 Fed. Reg. 24727 (June 11, 2025), https://www.federalregister.gov/documents/2025/06/11/2025-10814/unleashing-american-drone-dominance.

6
Advanced Manufacturing: Driving the Next Industrial Revolution

Elisabeth B. Reynolds

Strategic Importance

The importance of manufacturing capabilities and capacity to the US economy has become increasingly apparent in recent years as national and economic security issues related to US supply chains, innovation, and global competitiveness have become national priorities. Not only is manufacturing foundational to the industrial systems upon which the country operates (such as energy, transportation, or materials); it is also critical to the ability of the country to innovate and lead globally in strategically important technologies and industries, as the chapters in this volume underscore.

Advanced manufacturing is a term that brings manufacturing into the twenty-first century, moving beyond the antiquated vision of US manufacturing in its heyday in the mid twentieth century. It refers to a suite of manufacturing technologies as well as to production systems more broadly that are converging to create new manufactured products and processes. The technologies involve digitalizing the production process through automation, AI, sensors, robotics, and additive manufacturing, among other technologies, to improve quality, yields, safety, resilience, and overall productivity. Given the current shortage of manufacturing workers, implementation of these technologies in combination with organizational change to support digitalization and job redesign can also augment the capabilities and productivity of manufacturing workers.

A clear thread throughout all of the topics in this volume (critical minerals, semiconductors, biomanufacturing, quantum computing, and drones) is the important role of advanced manufacturing to each of these industries, either directly or indirectly. As previous MIT research has highlighted,[1] advanced manufacturing is essential to our ability to innovate and lead technologically in complex areas where hardware and software are highly integrated. It is also at the core of our ability to build the systems and industries that help drive economic growth and ensure national and economic security where some domestic manufacturing capacity is considered critical. Indeed, since 2022, the federal government has taken significant steps to invest in the domestic production of critical minerals, semiconductors, and drones.

Advances in manufacturing technologies are converging with a number of other factors to elevate production writ large as a national priority. First, the global pandemic exposed vulnerabilities in critical supply chains that were spread out globally and that prioritized efficiency (i.e., lower costs) over resilience. From personal protective equipment to semiconductors, supply chain bottlenecks put US national and economic security—including health security—at risk. Today, companies and governments are rethinking strategies to increase supply chain resilience, including through strengthening digital connectivity across suppliers to increase transparency, creating redundancy for key supplies, and localizing some production regionally.

Second, a changing geopolitical landscape and global economic order present growing threats to US national security interests. Advanced manufacturing capabilities and capacity are needed to strengthen the country's defense industrial base and ensure technological superiority over adversaries. The recent global conflict between

Russia and Ukraine and rising threats from China—technologically, economically, and militarily—have raised concerns about US preparedness for future conflicts as well as leadership in critical technologies where the US has fallen behind China.[2] All of the technologies discussed in this volume are dual use; that is, they can be used for both defense and civilian purposes, which can be beneficial to their commercialization and growth.

Third and last, a strong industrial base both relies on and can contribute to energy security, which is becoming a top priority in light of aforementioned challenges of increasing resilience, hedging against geopolitical conflict, and, more recently, ensuring a strong energy supply to support increasing energy demands from AI. Upgrading and enhancing the electric grid and developing and expanding new energy sources such as nuclear and geothermal require a strong industrial base. In addition, manufacturing companies are enhancing productivity by increasing energy efficiency, which can also lead to greater industrial decarbonization—a priority for many industrial companies as well as countries.

For all of the above reasons, we are at the beginning of what some refer to as "the next industrial revolution."[3] Advances in digital technology as well as prioritization of the factors listed here create the opportunity for more productive, resilient, and sustainable manufacturing in the twenty-first century. Importantly, we also have an opportunity to create quality jobs where investments in both technology and workers lead to increased productivity, higher wages, and—with the right labor market institutions in place—greater shared prosperity.[4] After decades of disinvestment in the US industrial base, there is general consensus that it is time to reindustrialize.

Current Landscape

Concerns about declining US manufacturing capacity and capabilities go back decades, as the repercussions from offshoring, globalization, the China Shock (the flood of exports from China into the United States after it entered the World Trade Organization in 2001),[5] and technological change have taken their toll on communities, workers, and national security writ large. The past four federal administrations have prioritized rebuilding the industrial base and introduced numerous programs and policies to try to turn around the decline as it relates to the scale, composition, and performance of US manufacturing.

As a percentage of the overall economy, the US manufacturing sector has declined in the past seventy years from more than 20 percent of US GDP in the 1950s to 10 percent today. By comparison, as of 2025, the manufacturing sector is approximately 25 percent of total GDP in China, 21 percent in Japan, and 18 percent in Germany.[6] Accordingly, US employment in manufacturing is low, at approximately 8 percent of the non-farm labor force (compared with 29, 19, and 16 percent in China, Germany, and Japan, respectively),[7] although manufacturing has one of the highest job multipliers (for every one job in manufacturing, seven jobs are created in related or supporting industries).[8]

Given these statistics, it is no surprise that the United States and China have reversed positions since 2010 in terms of percentage of global manufacturing output and value added, with China decidedly surpassing the United States (31 percent for China and 16 percent for the United States of the worldwide total).[9] China has invested heavily in its manufacturing capacity to become the manufacturing center of the world for both low value-added and higher value-added products. China's manufacturing capabilities have moved up the value chain over the years from production of textiles, toys, and furniture to more sophisticated products in key frontier technologies, including electric vehicles, batteries, solar, drones, and biomanufacturing, raising concerns not only about catching up to China but also about overcapacity and the risk of flooding global markets with Chinese products.[10]

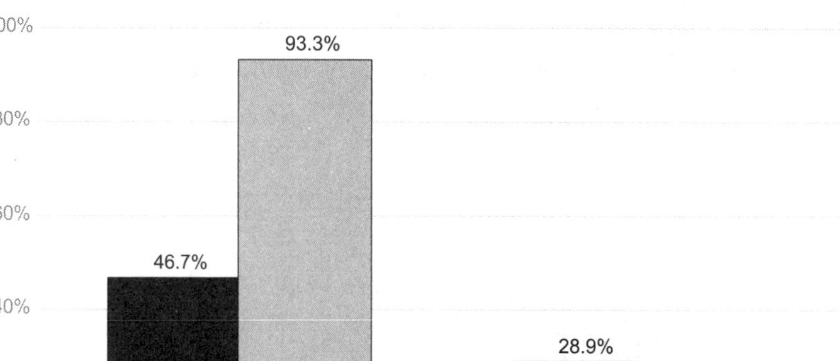

Figure 6.1
Cumulative growth in manufacturing and nonfarm business labor productivity (1991–2023). Data from the US Bureau of Labor Statistics, Nonfarm Business Sector: Labor Productivity (Output per Hour) for All Workers (OPHNFB); Manufacturing Sector: Labor Productivity (MFGOPH), retrieved from the Federal Reserve Bank of St. Louis (FRED). Chart adapted from Alex Muresianu, Tax Foundation.

More important than the overall output and employment numbers, however, is the rate of manufacturing productivity growth, particularly in higher value-added, advanced manufacturing industries where the United States needs to compete globally. Across all US manufacturing industries, manufacturing productivity has stagnated for over a decade, since roughly 2011 (see figure 6.1).[11] Without increases in productivity, manufacturing firms cannot increase their efficiency, improving output, skills, and wages in the process. According to some analyses, US manufacturing productivity lags behind many of its global competitors, including Taiwan, Germany, and South Korea (though not Japan).[12]

There are several explanations for the flatlining of manufacturing productivity in the United States, but a primary driver is the rate of technology adoption. Relative to other countries, US manufacturers underinvest in technology adoption. Only about 12 percent of American factories have any robotic automation, and only about one in three factories has specialized software and cloud computing capabilities. The United States ranks tenth in the world in terms of robot adoption density, behind Germany, Japan, and China.[13]

Related to low technology adoption are relatively low wages. The US manufacturing "wage premium" has declined from 40 percent in the 1950s to 4 percent in 2009 to –1 percent in 2019 (wages are higher in larger companies and for unionized jobs). As of 2025, manufacturing average hourly wages were below average private-sector wages, though higher in terms of weekly earnings because of steadier work hours.[14]

A critical area of competitive advantage for the United States in advanced manufacturing and a pathway to higher productivity is through its innovation capacity, in both products and processes. The private sector funds the majority of R&D in the United States (76 percent compared with 18 percent by the federal government), and three of the top five industries for private-sector R&D are manufacturing-related (chemical manufacturing, including pharmaceuticals; electronics and computers, including semiconductors; and transportation[15]). In 2021, US R&D "intensity" (combined public and private investments) trailed Taiwan and South Korea but surpassed Japan, Germany, and China.[16] Going forward, maintaining or increasing US R&D intensity will be important to achieve many of the goals outlined in this volume.

US strengths in software development position it well to lead in innovation in advanced manufacturing technologies. For example, the United States has the highest number of patents filed globally in additive manufacturing, with approximately 40 percent of international patents filed between 2001 and 2020, followed by Japan (13.9 percent) and Germany (13.4 percent). It also is a leader in specific robotics areas such as aerospace, medical, and robotics patents with AI features,[17] although China, Japan, and South Korea have led in the number of robotics patents overall.[18]

Historically, hardware-based startups have had a more difficult time scaling up in the United States due to the larger capital costs and longer timelines that are required to commercialize new technologies that require significant manufacturing and engineering expertise and equipment.[19] However, there is a new appreciation and interest in the United States in hardware-related startups (some of which fall into the category of "tough tech"), which has led to a recent focus by leading venture capital firms and other investors in the industrial sector. Whether this interest translates into more commercialized products and innovative advanced manufacturing firms of scale remains to be seen, but it is a source of strength for the US industrial base.[20]

Gaps and Opportunities

There are several areas where US manufacturing needs to be strengthened and expanded to support the long-term viability of the country's industrial base. These are described briefly below.

Technology Adoption

Despite leading in manufacturing technology innovation, relative to other countries, US manufacturers lag in technology adoption. US manufacturing productivity has been relatively flat since 2012,[21] in part because US manufacturers, particularly small- and medium-sized enterprises (SMEs), report less adoption of advanced manufacturing technologies such as cloud computing, robotics, and AI.[22] Without more accelerated digital technology adoption, US manufacturers, particularly SMEs, will find it challenging to become more productive, an important step for firms to offer higher wages to attract and retain workers. Digitalization is also critical to addressing supply chain resilience and industrial decarbonization.

Workforce Expansion and Skills Upgrading

Disinvestment and loss of manufacturing over several decades have led to a "lost generation" of manufacturing workers in the United States. Manufacturing employment fell by 33 percent between 1999 and 2010 but has since grown back since 2023 to roughly 2008 levels—although it remains far below 1999 levels. In the fall of 2025, there were approximately 400,000 open manufacturing jobs in the country, and it is predicted there will be close to two million unfilled jobs within a decade.[23]

Attracting workers into manufacturing jobs is challenging, given the history of deindustrialization and job decline. Rebuilding manufacturing talent in the country will take significant investment in national and state training institutions at all levels—vocational, community college, and higher education. Pre-apprenticeships and apprenticeships offer a promising path forward.[24] Manufacturing traditionally provided jobs with good career paths, and still does in some segments of manufacturing, but steps need to be taken to restore this status.[25] In addition to a supply-side strategy, the country also needs a demand-side strategy: research shows that 20–40 percent of companies that adopt advanced technologies report raising their demand for higher skills and training.[26] Such investments can lead to the redesigning of jobs to incorporate advanced technology and higher skills.

Manufacturing Scale-Up

Scaling innovative advanced manufacturing technologies from prototype to pilot and early-stage commercial production can require longer time horizons and larger capital expenditures—in the tens to hundreds of millions of dollars—than do pure software investments. These types of investments often have fewer investors, as they are less appealing to venture capitalists, who are willing to take risks but seek returns on a relatively short timeline, or to later-stage investors, who will accept lower returns and longer pay-back periods but only for relatively de-risked products. Thus, innovative tools and approaches are needed to help "pull" new manufacturing-related technologies through pilot and demonstration phases of scale-up and even large-scale deployment. This could involve innovative new private-sector mechanisms like "technology insurance" or government procurement mechanisms, or a combination of both.

Manufacturing Productivity

As highlighted above, manufacturing productivity—the amount of output per worker hour—has been flatlining in the United States since 2011. While it is complicated to tease out the exact factors at play, it is generally thought that a lack of investment in new technology is a primary driver of the stagnation. Increasing US productivity in manufacturing is essential for US leadership in key industries as well as for increasing wages and prosperity more broadly for US manufacturing workers.

Recommendations

1. Catalyze adoption of advanced manufacturing technologies and organizational practices to increase productivity and wages.

As outlined above, the United States generally lags in adoption of advanced manufacturing technologies relative to other regions of the world. This is particularly the case among SMEs, which represent the vast majority of US manufacturing establishments and employ approximately 40 percent of the manufacturing workforce. Despite the return on investment associated with digital technology adoption by SMEs, many are reluctant to invest because they are risk-adverse or short on capital. Without signals from customers and the marketplace more broadly, SMEs are reluctant to change existing manufacturing processes. Access to low-cost capital and financial incentives can help those SMEs that are interested in growing to make investments into digital manufacturing, including AI. However, more is needed than just technology acquisition in and of itself. A change in organizational mindset and culture with the associated changes in workforce skills and jobs is also considered essential to successful technology adoption. The following steps could be taken:

- Support state-led programs that offer incentives to SMEs willing to invest in new productivity-enhancing technologies that include hardware, software, and integration services. The Indiana Manufacturing Readiness Grant,[27] the Massachusetts Manufacturing Accelerate Program,[28] and the Michigan Industry 4.0 Technology Implementation Grants[29] are all examples of successful programs that can serve as models for programs in other states.
- Encourage large manufacturers that have contracts with or grants from the federal government to fully digitalize their supply chains. Federal agencies could offer incentives for original equipment manufacturers to invest in digitalization and better connectivity in their supply chains.
- Create a national, open-source platform of advanced digitalization tools and educational materials that can work at scale to encourage SMEs to learn about and adopt digital technologies and organizational practices in their operations. Leverage regional networks and the Manufacturing USA Institutes to elevate AI adoption among SMEs as a national priority. New or existing institutes and regional partners could undertake the following initiatives:
    - Create a library of highly accessible case studies to guide firms on how to identify the problems they want to solve and the appropriate tools for solving the problems, how to

integrate new technology into their operations, and how to create a continuous improvement mindset for digital manufacturing.

- ○ Forge partnerships and support educational programs at postsecondary institutions that will train the next generation of workers on digital manufacturing.
- ○ Leverage existing networks where centers of excellence in AI and manufacturing exist in university/industry consortia across the country.

2. Encourage investment strategies to accelerate manufacturing scale-up.

New financing models, tools, and entities are needed to address the "missing middle" capital gaps needed to scale advanced manufacturing technologies from pilot to demonstration-at-scale.[30] As underscored across the chapters in this volume, challenges in accessing patient capital exist across multiple industries including semiconductors, biomanufacturing, and advanced manufacturing technologies more broadly because of the scale of investments required over a sustained period of time where venture-level returns are harder to achieve and investors tend to prefer "asset light" investments. The following are some examples of strategies that can make such investments more compelling.

- *Use federal procurement to support commercialization of novel technologies.* Early-stage technologies struggle to secure customers because prices for their first few units may be too high, or customers may not be willing to commit until they have actually seen the products work. Many technological innovations and breakthroughs have come through the use of government procurement tools in areas such as health, defense, and energy. Tools such as advance market commitments and offtake agreements guarantee purchase of a product once it is brought to market. The US government used advance market commitments to develop the pneumococcal and COVID-19 vaccines through Operation Warp Speed.[31] Other Transaction Authorities are a tool that allow the federal government to bypass traditional procurement processes in order to engage non-traditional companies (i.e., startups) with the goal of supporting support innovative, early-stage product development or follow-on production. The US Department of Defense's Replicator drone initiative is one of many examples of the use of an Other Transaction Authority.[32] The federal government could increase its impact in priority areas by using these tools to accelerate innovation, technology adoption, and scale-up.

- *Explore new scale-up financing tools and entities* that could help scale production that requires a longer time horizon and increased capital expenditures for pilot and demonstration-at-scale production. Currently, US capital markets, specifically venture capital, are not well aligned with the long-term investments required to bring manufacturing/engineering-based startups to scale, given the amount of equity an investor must take early on to reach desirable returns. The federal government could take the following steps:
    - Engage the private sector in exploring new financing models. Some novel ideas include a "first of a kind" (FOAK) fund and "technology insurance" in which credit insurance is developed for lenders to cover performance risk on projects where traditional insurers are unlikely to appropriately price technology risk.[33]
    - Create a national investment fund like an Industrial Finance Corporation or a sovereign wealth fund to take a long-term view on strategic technology development and scale up.[34] Such an entity could help companies finance FOAK facilities and address gaps in production capacity in critical supply chains.

- Determine which existing financing vehicles and models across the federal government could be most useful and expanded for broader use in advanced manufacturing goals. Some models that could be examined for this purpose include BARDA Ventures, which invests in health security and innovative medical countermeasures, the Export-Import Bank's Make More in America program, the US Department of Defense's Office of Strategic Capital,[35] and the US Department of Energy's Loan Program Office.

- *Support the creation of shared "forge" facilities for pilot production* in critical technologies. Given the expense and limited use for any single company of building post-prototype facilities to test new technologies, a number of efforts are under way to support the US innovation engine by providing shared facilities for scale-up. These include the following:
  - *Semiconductors*: With the CHIPS Act, companies receiving funds from the US Department of Commerce to build foundries in the United States are required to create pathways for use by startups.[36]
  - *Biomanufacturing*: BioMADE, part of the Manufacturing USA network, with funding from the US Department of Defense, state governments, and the private sector, is investing $500 million to create a national Bioindustrial Pilot Plant Network.[37]
  - *Clean energy*: The US Department of Energy has supported the creation of multiple pilot facilities to develop geothermal and hydrogen capabilities, though some hydrogen hub funding was eliminated in 2025.

These models engage federal and state government as well as the private sector, providing useful models for addressing some of the barriers to early-stage scale-up at the pilot production phase.

3. Expand manufacturing workforce training and education. Extending and upskilling today's manufacturing workforce while attracting the next generation into new manufacturing jobs is a critical piece of the reindustrialization agenda in the United States. Currently, both the vocational training system and the community college system—the career and technical education system more broadly—need expanded and upgraded programs. This is already in process across the country, as seen in the expansion of programs in semiconductor training, and there are a number of successful programs upon which to build the country's manufacturing workforce capacity and capabilities. Several next steps could include the following:

- Create robust pre-apprenticeships and apprenticeship programs for manufacturing education that can be adapted from the Swiss and German models.[38] Connect apprenticeship programs across industry and community colleges.[39] Learn from existing state-wide programs.[40]
- Identify the most successful manufacturing training programs in the country and look to scale them.[41] The Department of Defense as well as a number of Manufacturing USA institutes[42] and community colleges have strong manufacturing training programs that could be scaled.
    - Bring advanced manufacturing into engineering curricula.[43]
    - Combine online education with learning-by-doing to scale advanced manufacturing workforce education.[44]
- Encourage four-year institutions of higher education to partner with community colleges, industry, and state and local governments to develop workforce programs, including for new manufacturing positions and career paths that fall between the roles of technician and engineer and that are needed to implement advanced manufacturing.[45]

- Fund research into evidence-based pedagogy and the application of learning sciences in order to improve manufacturing education and training.
- Gain a better understanding of which factors discourage individuals from going into manufacturing jobs, such as low wages, lack of awareness of good jobs or of internships and apprenticeships, stigma around blue-collar work, and lack of awareness of compelling technology.

4. Expand and deepen US manufacturing "ecosystem" capabilities through place-based investment strategies.

Today, there is a disconnect between US innovation capacity and its on-the-ground industrial capacity. As discussed in the biomanufacturing and drone chapters, the US will need to significantly expand its manufacturing capacity to meet both economic growth opportunities and national security objectives. To do this, the country must invest in its regional manufacturing ecosystems. US manufacturing capacity and capabilities are grounded in the context of regional economies and industry clusters that create both specialization and generalizable skills and capabilities.[46] Through the geographic concentration of suppliers, customers, workers, universities, startups, and institutions that help connect them with each other, a range of positive spillovers is created, including new knowledge and innovation, a skilled workforce, and specialized suppliers. Many successful regional manufacturing ecosystems have developed in the United States over the years[47]—automotive in Michigan and Ohio, aerospace and defense in Alabama and California, biomanufacturing in Massachusetts and North Carolina, and electronics and chemicals in Texas. Nearly every state has some manufacturing capacity; in some states such as Indiana, it accounts for close to a third of state GDP.[48]

Expanding and deepening US manufacturing ecosystems is arguably more about building networks and relationships to create synergies and spillovers than about large budgets. Along with developing a skilled workforce, regional manufacturing ecosystems can be supported through three key strategies outlined below.

- *SMM technology adoption.* As highlighted earlier, one of the major challenges facing US manufacturing is low productivity, particularly among small- and medium-sized manufacturers (SMM), due primarily to the challenges of adopting new manufacturing technologies and practices. The primary regional organizations that support SMMs in the United States have been grounded in adoption of lean manufacturing practices embodied by the Toyota Production System and introduced to the United States in the late 1980s through the Manufacturing Extension Partnership (MEP) system.[49] While a "lean mindset" continues to be essential for manufacturing, it must be integrated with the adoption of digital manufacturing capabilities, including AI. Intermediaries are needed[50] to help SMMs navigate a complex and dynamic technology market in which digital technology is now a prerequisite for competitive manufacturing performance. By marrying a lean mindset with forward-looking digital readiness, regional intermediaries can help SMMs improve productivity and upgrade skills. This may require creating a new set of metrics by which intermediaries are measured, one that emphasizes growth through innovation and technology adoption. The United States has a long history of working in this area upon which it can continue to build. Several countries also have strong programs from which the United States can learn, including Canada,[51] Germany (see its Fraunhofer Institutes and Industry 4.0 platform), and Israel.
- *Scale-up of early-stage advanced manufacturing technologies.* The United States has a competitive advantage in innovation in advanced manufacturing technologies but has faltered when it comes to their scale-up. To accelerate the move "from lab to

fab," the Manufacturing USA Innovation Institutes were developed just over a decade ago to facilitate scale-up from the lab to pilot production, where it is hoped the private or public sector would then pick up the scale-up of new products for commercialization. Modeled on the German Fraunhofer system, there are now eighteen Institutes overseen by the US Departments of Defense, Energy, and Commerce. They are valuable public/private efforts to advance the development and deployment of new manufacturing technologies and processes; however, they are in need of a revamping. The next iteration of the model needs to build on lessons learned to date and include the following initiatives.[52]

- Integrate the various centers into more of a network to help speed the adoption of new technologies that touch upon multiple Institutes.
- End provisions that sunset aid to the centers and instead make continued funding contingent on evaluations that align Institute work with national priorities and market opportunities.
- Extend Institute manufacturing readiness levels (MRLs) to include pilot production. The Institutes' initial mandate was to focus on MRLs 4–7, which fall between lab and pilot production. The Institutes should be leaning into pilot production and low-rate initial production to extend their MRL work to 8–9; this will help de-risk technologies further in order to attract private-sector investment. To date, it has been challenging to engage the private sector because products and processes are not sufficiently de-risked and advanced.
- Work with federal agencies and the private sector to develop "demand-pull" strategies to scale innovative products that are developed by the Institutes with their partners.

- *Regional manufacturing technology hubs.* The federal government has played a catalytic role over the years in helping places build upon their competitive advantages and develop their economic capacity, in particular industries and technologies. Recent initiatives look to expand the number of places that have world-class innovation and scientific strengths that could become global economic centers of excellence. With the Regional Technology and Innovation Hubs (Tech Hubs) initiative launched in 2023, the majority of the twelve designees directly or indirectly are developing advanced manufacturing capabilities in areas such as biomanufacturing, quantum computing, semiconductors, and polymers.[53] These investments are good for US manufacturing but also generate positive spillovers for the people, communities, and regions in which they are located. The federal government should continue to play this important role in developing US manufacturing ecosystems and could focus future investments in technology hubs on advanced manufacturing technologies and adoption. Making these investments in a strategic and coordinated fashion could lower risk and reduce the time of regional technological transformation from decades to years.

5. Increase federal R&D investments and target manufacturing processes.

Federal R&D funding plays a critical role in the US innovation engine because it generates positive spillovers for the country, compared with private-sector R&D investment (the private sector underinvests in R&D from a societal perspective), and also acts as a catalyst to private-sector investment, often fueling the development of technologies to a point at which they become more attractive from a commercial perspective.[54] Federal funding provides the majority of R&D expenditures on basic research (41 percent). If one includes university funding as well (13 percent), which is supported in part by federal dollars, the federal investment in R&D is over 50 percent. Basic research focuses on discovery, often providing the seeds to important applications decades before they emerge.[55]

As the chapters in this volume underscore, significant R&D is required in advanced manufacturing processes that drive breakthroughs in both science and engineering. Innovation in manufacturing processes, as much as in product innovation, can be an important source of productivity growth.[56] Whether breaking new ground in fields such as quantum computing or "leapfrogging" existing manufacturing processes in biomanufacturing, material sciences, semiconductors, robotics, or additive manufacturing, R&D investments in the basic science and engineering of advanced manufacturing technologies are essential to US national and economic security.

However, federal R&D funding has been trending in the wrong direction and, as of 2025, came under direct threat. Federal R&D expenditures have been relatively flat since the 1990s, growing at a compound annual growth rate of just over 1 percent between 1993 and 2023, compared with an average private-sector compound annual growth rate (CAGR) of approximately 5 percent.[57] Recent steps to significantly cut federal R&D funding, including to universities where much of the country's innovative research occurs,[58] will be counterproductive to US goals to lead globally in key frontier technologies, so many of which require advanced manufacturing capabilities.

6. Use trade tools selectively to ensure national security priorities and fair-trade practices.

Tariffs, which are a tax on imports, should be used strategically to promote and protect US manufacturing in situations involving either a clear national security interest or unfair trade practices that undermine US industries. In recent years, US administrations from both political parties have used tariffs for both situations, most often focused on Chinese advancement in frontier technologies as well as on Chinese overproduction and the unfair dumping of products on US markets.[59] However, in the past, tariffs were used as a scalpel, imposed on specific countries, industries, and technologies of concern. With the Trump administration, they have been deployed as a hatchet—broad-based tariffs levied across a wide range of products on friends and foes alike.

Across-the-board tariffs are particularly challenging for manufacturing because they impact not just final products but the intermediate inputs to all types of products (e.g., automobiles, computers and electronics, machinery, chemicals, and transportation). While tariffs might be absorbed for a period of time by companies, ultimately their cost is passed on to customers, driving higher prices, lower employment, and ultimately slower economic growth.[60]

Not all manufacturing is equal. The United States will have difficulty competing globally in lower value-added manufactured products (such as clothing and toys), including commodities, if the manufacturing cannot support relatively high US wages. The US competitive advantage is found in higher value-added manufacturing where innovation, technology, and skills are rewarded by the marketplace. At the margins, tariffs might lead to some new manufacturing investments in the United States, but in general, increased costs resulting from tariffs will put US manufacturing at a global disadvantage. US policies should be strategic, focusing on industries that are critical to US national and economic security, in terms of both frontier technologies and supply chain resilience, and should create incentives to encourage investments in increased capabilities and capacity.

Conclusion

A collective recognition of the importance of manufacturing to the US economy is converging with new technologies to offer a new chapter and reinvigoration of the US industrial base. This new trajectory—one in which manufacturing can be more productive, resilient, and sustainable—speaks to both the foundational industrial systems upon which the economy is built and the innovative frontier industries and technologies that are discussed in this volume. By accelerating technology development and adoption, the United States can both defend its national security interests and lead globally in growth industries that can drive greater prosperity. Importantly, by improving the quality of workforce training in the age of AI as well as the general quality of manufacturing jobs, which on average are lagging in wages compared with other private-sector jobs, manufacturing can support shared prosperity goals. While reindustrializing the country will not translate into the scale of employment that existed in the 1970s, it can translate into building more career pathways to the middle class and economically more robust communities and regions.

However, this industrial future faces several hurdles related to technology adoption (particularly by SMMs), workforce development, the scaling of new manufacturing technologies, and R&D investments in "leapfrog" technologies. In addition, the scale at which the United States needs to expand its manufacturing capacity will require concerted efforts at the regional level to build out manufacturing ecosystems that benefit from the positive spillovers that come from a well-connected network and public/private partnerships.

The author would like to thank the following MIT colleagues for their contribution to this chapter: Ben Armstrong, Executive Director, Industrial Performance Center, Co-Lead, Work of the Future Initiative; Suzanne Berger, Institute Professor, Department of Political Science; William Bonvillian, Lecturer in the Science Technology and Society and Political Science Departments; Julie Diop, Executive Director, MIT Initiative for New Manufacturing; John Hart, Class of 1922 Professor, Head, Department of Mechanical Engineering, Director, Laboratory for Manufacturing and Productivity,

Director, Center for Advanced Production Technologies, and Co-Director, MIT Initiative for New Manufacturing; John Liu, Principal Investigator, MIT Learning Engineering and Practice (LEAP) Group; J. Christopher Love, Raymond A. (1921) and Helen E. St. Laurent Professor of Chemical Engineering, Associate Director of the Koch Institute for Integrative Cancer Research at MIT, Co-Chair of MIT's Initiative for New Manufacturing (MIT INM), Co-Director of the Leader for Global Operations (MIT LGO); and David Mindell, Frances and David Dibner Professor of the History of Engineering and Manufacturing (STS) Professor of Aeronautics and Astronautics.

Notes

1. Suzanne Berger, *Making in America: From Innovation to Market* (MIT Press, 2013).

2. Department of Defense, *National Defense Industrial Strategy* (2023), https://www.businessdefense.gov/docs/ndis/2023-NDIS.pdf; Australian Strategic Policy Institute, "ASPI's Two-Decade Critical Technology Tracker: The Rewards of Long-Term Research Investment," August 28, 2024, https://www.aspi.org.au/report/aspis-two-decade-critical-technology-tracker/.

3. David A. Mindell, *The New Lunar Society: An Enlightenment Guide to the Next Industrial Revolution* (MIT Press, 2025).

4. David Autor, David Mindell, and Elisabeth Reynolds, *The Work of the Future: Building Better Jobs in an Age of Intelligent Machines* (MIT Press, 2022). Labor market institutions can refer to a range of policies, including minimum wage, unemployment and wage insurance, and career and technical education.

5. David H. Autor, David Dorn, and Gordon H. Hanson. "The China Shock: Learning from Labor-Market Adjustment to Large Changes in Trade," *Annual Review of Economics* 8, no. 1 (2016): 205–240.

6. World Bank, "Manufacturing, Value Added (% of GDP)," accessed July 22, 2025, https://data.worldbank.org/indicator/NV.IND.MANF.ZS.

7. "Data Page: Manufacturing Jobs as a Share of Total Employment," *Our World in Data*, 2025, https://archive.ourworldindata.org/20250624-125417/grapher/manufacturing-share-of-total-employment.html.

8. Josh Bivens, "Updated Employment Multipliers for the U.S. Economy," Economic Policy Institute, January 23, 2019, https://www.epi.org/publication/updated-employment-multipliers-for-the-u-s-economy/.

9. Douglas Thomas, *Annual Report on the U.S. Manufacturing Economy: 2024*, NIST Advanced Manufacturing Series 600-16 (Gaithersburg, MD: National Institute of Standards and Technology, 2024), https://doi.org/10.6028/NIST.AMS.600-16; Mark J. Perry, "Chart of the Day: China Is Now World's No. 1 Manufacturer," American Enterprise Institute, December 12, 2012, https://www.aei.org/carpe-diem/chart-of-the-day-china-is-now-worlds-no-1-manufacturer/.

10. See James McBride and Andrew Chatzky, "Is 'Made in China 2025' a Threat to Global Trade?," Council on Foreign Relations, May 13, 2019; Evelyn Cheng, "China Doubles Down on Manufacturing, Leaving Real Estate Behind," *CNBC*, March 8, 2024; Nathaniel Taplin, "Why China's Overcapacity Problem Is About to Get Even Worse, in Seven Charts," *Wall Street Journal*, June 4, 2024.

11. Noah Smith, "Why Has U.S. Manufacturing Productivity Stagnated for Over a Decade?," *Noahpinion* (blog), July 6, 2024, https://www.noahpinion.blog/p/why-has-us-manufacturing-productivity.

12. Greg Ip, "American Labor's Real Problem: It Isn't Productive Enough," *Wall Street Journal*, September 20, 2023, https://www.wsj.com/economy/american-labors-real-problem-it-isnt-productive-enough-185fb9f1.

13. Ben Armstrong and Julie Shah, "A Smarter Strategy for Using Robots," *Harvard Business Review* 101, nos. 3–4 (2023): 35–42; see also "U.S. Manufacturing Economy," NIST, accessed July 22, 2025, https://www.nist.gov/el/applied-economics-office/manufacturing/manufacturing-economy/total-us-manufacturing.

14. David Autor, presentation at the Manufacturing@MIT symposium, May 2023; U.S. Bureau of Labor Statistics, table B-3, "Average hourly and weekly earnings of all employees on private nonfarm payrolls by industry sector, seasonally adjusted," August 2025, https://www.bls.gov/news.release/empsit.t19.htm.

15. Five industries represent approximately 80 percent of all private-sector investment in R&D (a total of $603 billion in 2022): information (including software publishing) at 25 percent; chemicals manufacturing (including pharmaceuticals and medicines) at 18 percent; computer and electronic products manufacturing (including semiconductors) at 17 percent; professional, scientific, and technical services at 11 percent; and transportation equipment manufacturing (including auto and aerospace products and parts) at 8 percent. See National Science Board, National Science Foundation, *Research and Development: U.S. Trends and International Comparisons*, Science and Engineering Indicators 2024, NSB-2024-6 (National Science Foundation, 2024), https://ncses.nsf.gov/pubs/nsb20246/.

16. National Science Board, National Science Foundation, *Science and Engineering Indicators 2024: The State of U.S. Science and Engineering*, NSB-2024-3 (National Science Foundation, 2024), https://ncses.nsf.gov/pubs/nsb20243.

17. For patents in additive manufacturing, see European Patent Office, "Patent Filings in 3D Printing Grew Eight Times Faster than Average of All Technologies in Last Decade," September 19, 2023, https://www.epo.org/en/news-events/news/patent-filings-3d-printing-grew-eight-times-faster-average-all-technologies-last; for robotics patenting globally, see Margarita Konaev and Sara M. Abdulla, "Trends in Robotics Patents: A Global Overview and an Assessment of Russia," Center for Security and Emerging Technology, November 2021, https://doi.org/10.51593/20210012.

18. Konaev and Abdulla, "Trends in Robotics Patents."

19. Elisabeth B. Reynolds, Hiram M. Samel, and Joyce Lawrence, "Learning by Building: Complementary Assets and the Migration of Capabilities in US Innovative Firms," *Production in the Innovation Economy* (2014): 51–80.

20. Farah Stockman, "Two Days Inside the Movement to Reindustrialize and Rearm America," *New York Times*, July 22, 2025, https://www.nytimes.com/2025/07/20/us/manufacturing-tech-trump-reindustrialize.html.

21. State Science & Technology Institute, "Useful Stats: Is US Manufacturing Productivity on a Decline? A Detailed Look at BLS OPT Data," November 2, 2023, https://ssti.org/blog/useful-stats-us-manufacturing-productivity-decline-detailed-look-bls-opt-data.

22. See Daron Acemoglu et al., *Automation and the Workforce: A Firm-Level View from the 2019 Annual Business Survey*, NBER Working Paper 30659 (National Bureau of Economic Research, 2022). In 2015, the McKinsey Global Institute noted that achieving Industry 4.0 will require upgrading about 40–50 percent of the current asset base across US manufacturing industries; see Sree Ramaswamy et al., *Making It In America: Revitalizing U.S. Manufacturing* (McKinsey Global Institute, November 2017), https://www.mckinsey.com/~/media/mckinsey/featured%20insights/americas/making%20it%20in%20america%20revitalizing%20us%20manufacturing/making-it-in-america-revitalizing-us-manufacturing-full-report.pdf.

23. National Association of Manufacturers, "Facts About Manufacturing," NAM, accessed July 22, 2025, https://nam.org/mfgdata/facts-about-manufacturing-expanded/.

24. Adeola Lawal, Charlie Meynet, and Krizia Lopez, "Expanding Youth Apprenticeships," Harvard Kennedy School, August 2023, https://www.pw.hks.harvard.edu/post/expanding-youth-apprenticeships; William B. Bonvillian and Sanjay E. Sarma, *Workforce Education: A New Roadmap* (MIT Press, 2021).

25. Armstrong and Shah, "Smarter Strategy."

26. Acemoglu et al., *Automation and the Workforce.*

27. "Manufacturing Readiness Grants," Conexus Indiana, accessed August 1, 2025, https://www.conexusindiana.com/drive-industry-success/manufacturing-readiness-grants-technology-adoption/.

28. "Massachusetts Manufacturing Accelerate Program (MMAP)," Massachusetts Center for Advanced Manufacturing (CAM), accessed August 1, 2025, https://cam.masstech.org/mmap.

29. "Industry 4.0 Technology Implementation Grant," Michigan Economic Development Corporation, accessed August 1, 2025, https://www.michiganbusiness.org/industry4-0/grant/.

30. For an overview of the challenge as it relates to the clean energy industry, see "The Missing Middle: Capital Imbalances in the Energy Transition," SG2 Ventures, September 8, 2023, https://www.s2ginvestments.com/insights/missing-middle. Earlier research at MIT highlighted the challenge across a number of technology areas: Elisabeth B. Reynolds, Hiram M. Samel, and Joyce Lawrence, "Learning by Building: Complementary Assets and the Migration of Capabilities in U.S. Innovative Firms," in *Production in the Innovation Economy*, ed. Richard M. Locke and Rachel L. Wellhausen (MIT Press, 2014), 81–107, https://academic.oup.com/mit-press-scholarship-online/book/18310/chapter/176337624.

31. Renaissance Philanthropy, "Market Shaping," accessed July 23, 2025, https://renaissancephilanthropy.org/playbooks/market-shaping/.

32. William C. Greenwalt, "DOD's Replicator Program: Challenges and Opportunities," American Enterprise Institute, October 19, 2023, https://docs.house.gov/meetings/AS/AS35/20231019/116484/HHRG-118-AS35-Wstate-GreenwaltW-20231019.pdf.

33. Yuqi Zhu and Aaron Bergman, *Policies for Building the Technology Performance Insurance Market*, Report 25-03 (Washington, DC: Resources for the Future, February 2025), https://media.rff.org/documents/Report_25-03_e3BbFMc.pdf.

34. "Summary Industrial Finance Corporation Act of 2021," https://www.coons.senate.gov/imo/media/doc/SUMMARY%20IFCUS%20117%20v.2.pdf; Adnan Mazarei, Anna Gelperin and Edwin Truman, "A US sovereign wealth fund? A confused solution to an undefined problem," Peterson Institute for International Economics, February 12, 2025, https://www.piie.com/blogs/realtime-economics/2025/us-sovereign-wealth-fund-confused-solution-undefined-problem.

35. At the time of this writing, the DoD Office of Strategic Capital received close to $1 billion in funding in the most recent reconciliation bill passed by Congress in July 2025.

36. Arrian Ebrahimi and Jordan Schneider, "How to Make NSTC a Moonshot Success," Institute for Progress, April 22, 2024, https://ifp.org/how-to-make-the-nstc-a-moonshot-success/.

37. BioMADE, "BioMADE Names Six States in Review for Expansion of Bioindustrial Manufacturing Infrastructure Network," March 13, 2024, https://www.biomade.org/news/biomade-names-six-states-in-review-for-expansion-ofnbsp-nbspbioindustrial-manufacturing-infrastructure-networknbsp.

38. See Lawal et al., "Expanding Youth Apprenticeships"; Bonvillian and Sarma, *Workforce Education.*

39. "Virtual Apprenticeship Network Home," American Association of Community Colleges, accessed July 22, 2025, https://www.aacc.nche.edu/programs/workforce-economic-development/expanding-community-college-apprenticeships/intro-virtual-apprenticeship-network/.

40. David Tobenkin, "Indiana Leans into Youth Apprenticeships," *Community College Daily,* March 5, 2024, https://www.ccdaily.com/2024/03/indiana-leans-into-youth-apprenticeships/; "Apprenticeship Carolina," Apprenticeship Carolina, accessed July 22, 2025, https://www.apprenticeshipcarolina.com.

41. Bonvillian and Sarma, *Workforce Education.*

42. William Bonvillian, *The Playbook—For Workforce Education at Manufacturing Innovation Institutes,* January 2022, http://dx.doi.org/10.13140/RG.2.2.36521.88166.

43. "Strengthening the Talent for National Defense: Infusing Advanced Manufacturing in Engineering Education Through Capstone Design Courses," National Academies, accessed July 22, 2025, https://www.nationalacademies.org/our-work/strengthening-the-talent-for-national-defense-infusing-advanced-manufacturing-in-engineering-education-through-capstone-design-courses.

44. Bonvillian and Sarma, *Workforce Education.*

45. John Liu and William B. Bonvillian, "The Technologist," *Issues in Science and Technology* 40, no. 2 (Winter 2024): 43–48, https://doi.org/10.58875/TNFP2942.

46. Mercedes Delgado, Michael E. Porter, and Scott Stern, "Clusters, Convergence, and Economic Performance," *Research Policy* 43, no. 10 (2014): 1785–1799.

47. Martin Neil Baily and Nicholas Montalbano, "Clusters and Innovation Districts: Lessons from the United States Experience," Brookings Institution, May 2018, https://www.brookings.edu/wp-content/uploads/2018/05/es_20180508_bailyclustersandinnovation.pdf.

48. National Association of Manufacturers, "Manufacturing's Share of Gross State Product," accessed July 22, 2025, https://www.nam.org/wp-content/uploads/2019/05/MFG-GSP-FactSheet_201810.pdf.

49. The Toyota Production System's lean manufacturing practices were introduced into the United States, first through the automotive industry. Subsequently, the Manufacturing Extension Partnerships, introduced by the federal government in 1998, have played an important role across states to introduce lean concepts to SMMs.

50. Multiple models exist in other countries that combine public funding with private-sector or nonprofit implementation. See Robert D. Atkinson, *Accelerating Digital Technology Adoption Among U.S. Small and Medium-Sized Manufacturers*, Information Technology and Innovation Foundation, 2024, https://itif.org/publications/2024/04/19/accelerating-digital-technology-adoption-among-smes/.

51. Canada recently invested $100 million in the National Research Council's Industry Research Assistance Program (IRAP) to help small- and medium-sized businesses scale up and increase productivity by building and deploying new AI solutions (2024). See Atkinson, *Accelerating Digital Technology Adoption.*

52. For more detail, see David Adler and William Bonvillian, "America's Advanced Manufacturing Problem—And How to Fix It," *American Affairs* 7, no. 3 (2023).

53. "Regional Technology and Innovation Hubs (Tech Hubs)," US Economic Development Administration, accessed August 1, 2025, https://www.eda.gov/funding/programs/regional-technology-and-innovation-hubs.

54. Robert D. Atkinson, "Why Federal R&D Policy Needs to Prioritize Productivity to Drive Growth and Reduce the Debt-to-GDP Ratio," Information Technology and Innovation Foundation, September 12, 2019, https://itif.org/publications/2019/09/12/why-federal-rd-policy-needs-prioritize-productivity-drive-growth-and-reduce/.

55. M. Anthony Mills, "Why the Federal Government Must Put More Money Toward Basic Science," *The Hill*, June 5, 2021, https://thehill.com/opinion/finance/551730-why-the-federal-government-must-put-more-money-toward-basic-science/.

56. Mills, "Why the Federal Government Must Put More Money."

57. National Science Board, National Science Foundation, *Trends in U.S. R&D Performance and Funding*, Science and Engineering Indicators 2025, NSB-2025-7 (National Science Foundation, 2025), https://ncses.nsf.gov/pubs/nsb20257/trends-in-u-s-r-d-performance-and-funding.

58. "Trump's Proposed Budget Would Mean 'Disastrous' Cuts to Science," *Science*, May 2, 2025, https://www.science.org/content/article/trump-s-proposed-budget-would-mean-disastrous-cuts-science; L. Rafael Reif, "America's Coming Brain Drain," *Foreign Affairs*, May 6, 2025, https://www.foreignaffairs.com/united-states/trump-universities-war-america-coming-brain-drain.

59. Chad P. Bown, "US–China Trade War Tariffs: An Up-to-Date Chart," Peterson Institute for International Economics, August 1, 2025, https://www.piie.com/research/piie-charts/2019/us-china-trade-war-tariffs-date-chart.

60. Noah Smith, "Why Tariffs Haven't Raised Inflation Much (Yet)," *Noahpinion* (blog), July 11, 2025, https://www.noahpinion.blog/p/why-tariffs-havent-raised-inflation.

Acknowledgments

This volume is the product of a rich and vibrant exchange among a community of MIT faculty, researchers, and students as well as invited participants and guests who took part in the Priority Technologies Seminar between 2023 and 2025. The seminar was cochaired by Simon Johnson and Elisabeth Reynolds, with additional leadership provided by Brian Deese, MIT Innovation Fellow, who encouraged the seminar participants to focus on strategies for the country that speak to the near-term. The papers benefited from review and input from over forty members of the MIT community, representing the enduring spirit of MIT as an institution dedicated to applying research to produce practical benefits for society and a strong commitment to service to the country. David Goldston, Director of MIT's Washington, DC office until the spring of 2025, provided excellent guidance and editorial advice. Richa Vera Udayana provided outstanding research assistance, and Lisa Pinto provided stellar copyediting support. Michelle Fiorenza, always positive and ready to help, ensured the seminar was well managed.

Contributors

Jesús A. del Alamo is Donner Professor and Professor of Electrical Engineering at MIT. He is a member of the Royal Spanish Academy of Engineering and a Fellow of the Institute of Electrical and Electronics Engineers, the American Physical Society, and the Materials Research Society. He has worked at the forefront of research and education in semiconductor technology for his entire career. Among other awards, he has received Intel's Outstanding Researcher Award, the University Researcher Award by the Semiconductor Industry Association and the Semiconductor Research Corporation, and the Technical Excellence Award by the Semiconductor Research Corporation.

Simon Johnson is the Ronald A. Kurtz (1954) Professor of Entrepreneurship at the MIT Sloan School of Management, where he is head of the Global Economics and Management group. At MIT, he is co-director of the Stone Center on Inequality and Shaping the Future of Work. In 2024, Johnson received the Nobel Prize in Economics with Daron Acemoglu and James A. Robinson, "for studies of how institutions are formed and affect prosperity."

J. Christopher Love is the Raymond A. (1921) and Helen E. St. Laurent Professor of Chemical Engineering, and Associate Director of the Koch Institute for Integrative Cancer Research at MIT. He is a co-chair of MIT's Initiative for New Manufacturing (MIT INM) and the Co-Director of the Leader for Global Operations (MIT LGO) for the MIT School of Engineering. He is a serial entrepreneur who has cofounded multiple companies in life science technologies, diagnostics, and biomanufacturing.

Fiona E. Murray is the William Porter (1967) Professor of Entrepreneurship at the MIT School of Management. She serves as Chair of the NATO Innovation Fund. Her research examines the intersection of science, innovation, and industrialization for critical technologies in light of current geopolitical shifts. Her MIT Press book *Accelerating Innovation: Competitive Advantage Through Ecosystem Engagement* (2025) examines the role of innovation ecosystems in supporting governments and corporations in meeting their technology priorities.

William D. Oliver is the Henry Ellis Warren (1894) Professor of Electrical Engineering and Computer Science and a Professor of Physics at MIT. His research addresses the materials growth, fabrication, design, and measurement of superconducting qubits, as well as the development of cryogenic packaging and control electronics. From 2003 to 2023, he also worked at MIT Lincoln Laboratory. He is a Fellow of the American Association for the Advancement of Science, a Fellow of the American Physical Society, and a Fellow of the IEEE. He serves on the US National Quantum Initiative Advisory Committee.

Elsa A. Olivetti is the Jerry McAfee (1940) Professor of Engineering, Department of Materials Science and Engineering. Her work focuses on reducing the burden of materials production and consumption through increased use of recycled and waste materials; informing the early-stage design of new materials for effective scale-up; and understanding the implications of policy, new technology development, and manufacturing processes on materials supply chains.

Elisabeth B. Reynolds is Professor of the Practice in MIT's Department of Urban Studies and Planning. She was Special Assistant to the President for Manufacturing and Economic Development at the National Economic Council in 2021–2022. She co-led MIT's Task Force on the Work of the Future while executive director of the MIT Industrial Performance Center between 2010 and 2021.

Jonathan Ruane is a Senior Lecturer in Global Economics and Management at MIT Sloan and a Research Scientist at MIT's Initiative on the Digital Economy. He is a Principal Investigator and leads Thrust 4 (societal impacts) research for the NSF's Centre for Quantum Networks (CQN).

Index

Page locators in italics indicate figures.